Cohomological Invariants: Exceptional Groups and Spin Groups

MEMOIRS
of the
American Mathematical Society

Number 937

Cohomological Invariants: Exceptional Groups and Spin Groups

Skip Garibaldi
(with an Appendix by Detlev W. Hoffmann)

July 2009 • Volume 200 • Number 937 (second of 6 numbers) • ISSN 0065-9266

American Mathematical Society
Providence, Rhode Island

2000 *Mathematics Subject Classification.*
Primary 11E72; Secondary 12G05, 20G15, 17B25.

Library of Congress Cataloging-in-Publication Data

Garibaldi, Skip, 1972–
Cohomological invariants : exceptional groups and spin groups / Skip Garibaldi ; with an appendix by Detlev W. Hoffmann.
p. cm. — (Memoirs of the American Mathematical Society, ISSN 0065-9266 ; no. 937)
"Volume 200, number 937 (second of 6 numbers)."
Includes bibliographical references and index.
ISBN 978-0-8218-4404-5 (alk. paper)
1. Galois cohomology. 2. Linear algebraic groups. 3. Invariants. I. Title.

QA169.G367 2009
515′.23—dc22 2009008059

Memoirs of the American Mathematical Society

This journal is devoted entirely to research in pure and applied mathematics.

Subscription information. The 2009 subscription begins with volume 197 and consists of six mailings, each containing one or more numbers. Subscription prices for 2009 are US$709 list, US$567 institutional member. A late charge of 10% of the subscription price will be imposed on orders received from nonmembers after January 1 of the subscription year. Subscribers outside the United States and India must pay a postage surcharge of US$65; subscribers in India must pay a postage surcharge of US$95. Expedited delivery to destinations in North America US$57; elsewhere US$160. Each number may be ordered separately; *please specify number* when ordering an individual number. For prices and titles of recently released numbers, see the New Publications sections of the *Notices of the American Mathematical Society.*

Back number information. For back issues see the *AMS Catalog of Publications.*

Subscriptions and orders should be addressed to the American Mathematical Society, P. O. Box 845904, Boston, MA 02284-5904, USA. *All orders must be accompanied by payment.* Other correspondence should be addressed to 201 Charles Street, Providence, RI 02904-2294, USA.

Memoirs of the American Mathematical Society (ISSN 0065-9266) is published bimonthly (each volume consisting usually of more than one number) by the American Mathematical Society at 201 Charles Street, Providence, RI 02904-2294, USA. Periodicals postage paid at Providence, RI. Postmaster: Send address changes to Memoirs, American Mathematical Society, 201 Charles Street, Providence, RI 02904-2294, USA.

This publication is indexed in *Science Citation Index®*, *SciSearch®*, *Research Alert®*, *CompuMath Citation Index®*, *Current Contents®/Physical, Chemical & Earth Sciences.*
Printed in the United States of America.

∞ The paper used in this book is acid-free and falls within the guidelines established to ensure permanence and durability.
Visit the AMS home page at `http://www.ams.org/`

10 9 8 7 6 5 4 3 2 1 14 13 12 11 10 09

Contents

Abstract

This volume concerns invariants of G-torsors with values in mod p Galois cohomology — in the sense of Serre's lectures in the book *Cohomological invariants in Galois cohomology* — for various simple algebraic groups G and primes p. We determine the invariants for the exceptional groups F_4 mod 3, simply connected E_6 mod 3, E_7 mod 3, and E_8 mod 5. We also determine the invariants of Spin_n mod 2 for $n \leq 12$ and construct some invariants of Spin_{14}.

Along the way, we prove that certain maps in nonabelian cohomology are surjective. These surjectivities give as corollaries Pfister's results on 10- and 12-dimensional quadratic forms and Rost's theorem on 14-dimensional quadratic forms. This material on quadratic forms and invariants of Spin_n is based on unpublished work of Markus Rost.

An appendix by Detlev Hoffmann proves a generalization of the Common Slot Theorem for 2-Pfister quadratic forms.

Received by the editor September 14, 2006, and in revised form April 20, 2007.

2000 *Mathematics Subject Classification.* Primary 11E72; Secondary 12G05, 20G15, 17B25.

Key words and phrases. cohomological invariant, exceptional group, spin group, Rost's theorem.

List of Tables

Preface

These notes are divided into three parts.

The first part (§§1–8) is based on material originally developed for inclusion in Serre's lecture notes in [**GMS**]. (In particular, §4 on the Bockstein has hardly been changed at all.) The first part of the notes culminates with the determination of the invariants of PGL_p mod p (for p prime) and the invariants of Albert algebras (equivalently, groups of type F_4) mod 3.

The second part (§§9–15) describes a general recipe for finding a subgroup N of a given semisimple group G such that the natural map $H^1_{\text{fppf}}(*, N) \to H^1(*, G)$ is surjective. It is a combination of two ideas: that parabolic subgroups lead to representations with open orbits, and that such representations lead to surjective maps in cohomology. I learned the second idea from Rost [**Rost 99b**], but both ideas seem to have been discovered and re-discovered many times. Representation theorists will note that our computations of stabilizers N for various G and V—summarized in Table 23a—are somewhat more precise than the published tables, in that we compute full stabilizers and not just identity components. The surjectivities in cohomology are used to describe the mod 3 invariants of the simply connected split E_6 and split E_7's.

The last two sections of this part, 14 and 15, concern groups of type E_8 relative to the prime 5. The two main results in Section 14, namely Prop. 14.7 and Th. 14.14, describe a construction of groups of type E_8 and how it relates to the Rost invariant; they are contained in Chernousov's paper [**Ch 95**]. The proof of Th. 14.14 given here is different and uses arguments and results from Gille's papers [**Gi 00**] and [**Gi 02a**].

The third part (§§16–23) describes the mod 2 invariants of the groups Spin_n for $n \leq 12$ and $n = 14$. It may be viewed as a fleshed-out version of Rost's unpublished notes [**Rost 99b**] and [**Rost 99c**]. A highlight of this part is Rost's Theorem 21.3 on 14-dimensional quadratic forms in I^3.

There are also two appendices. The first uses cohomological invariants to give new examples of anisotropic groups of types E_7, answering a question posed by Zainoulline. The second appendix—written by Hoffmann—proves a generalization of the "common slot theorem" for 2-Pfister quadratic forms. This result is used to construct invariants of Spin_{12} in §20.

These are notes for a series of talks I gave in a "mini-cours" at the Université d'Artois in Lens, France, in June 2006. Consequently, some material has been included in the form of exercises. Although this is a convenient device to avoid going into tangential details, no substantial difficulties are hidden in this way. The exercises are typically of the "warm up" variety. On the other end of the spectrum, I have included several open problems. "Questions" lie somewhere in between.

Acknowledgements. It is a pleasure to thank Jean-Pierre Serre and Markus Rost, both for their generosity with their results and for their comments on this note; Detlev Hoffmann for providing Appendix B; and Pasquale Mammone for his hospitality during my stay in Lens. Gary Seitz and Philippe Gille both gave helpful answers to questions. I thank also Raman Parimala, Zinovy Reichstein, Adrian Wadsworth, and the referee for their comments.

Part I

Invariants, especially modulo an odd prime

1. Definitions and notations

1.1. DEFINITION OF COHOMOLOGICAL INVARIANT. We assume some familiarity with the notes from Serre's lectures from [**GMS**], which we refer to hereafter as S.[a] A reader seeking a more leisurely introduction to the notion of invariants should see pages 7–11 of those notes.

We fix a base field k_0 and consider functors

$$A\colon \mathsf{Fields}_{/k_0} \to \mathsf{Sets}$$

and

$$H\colon \mathsf{Fields}_{/k_0} \to \mathsf{Abelian\ Groups},$$

where $\mathsf{Fields}_{/k_0}$ denotes the category of field extensions of k_0. In practice, $A(k)$ will be the Galois cohomology set $H^1(k, G)$ for G a linear algebraic group[b] over k_0. In S, various functors H were considered (e.g., the Witt group), but here we only consider abelian Galois cohomology.

An *invariant of A (with values in H)* is a morphism of functors $a: A \to H$, where we view H as a functor with values in Sets. Unwinding the definition, an invariant of A is a collection of functions $a_k : A(k) \to H(k)$, one for each $k \in \mathsf{Fields}_{/k_0}$, such that for each morphism $\phi\colon k \to k'$ in $\mathsf{Fields}_{/k_0}$, the diagram

$$\begin{array}{ccc} A(k) & \xrightarrow{a_k} & H(k) \\ {\scriptstyle A(\phi)}\downarrow & & \downarrow{\scriptstyle H(\phi)} \\ A(k') & \xrightarrow{a_{k'}} & H(k') \end{array}$$

commutes.

1.2. EXAMPLES. (1) Fix a natural number n and write $\mathscr{S}_n$ for the symmetric group on n letters. The set $H^1(k, \mathscr{S}_n)$ classifies étale k-algebras of dimension n up to isomorphism. The sign map $\mathrm{sgn}\colon \mathscr{S}_n \to \mathbb{Z}/2\mathbb{Z}$ is a homomorphism of algebraic groups and so defines a morphism of functors—an invariant—$\underline{\mathrm{sgn}}\colon H^1(*, \mathscr{S}_n) \to H^1(*, \mathbb{Z}/2\mathbb{Z})$. The set $H^1(k, \mathbb{Z}/2\mathbb{Z})$ classifies quadratic étale k-algebras, i.e., separable quadratic field extensions together with the trivial class corresponding to $k \times k$, and $\underline{\mathrm{sgn}}$ sends a dimension n algebra to its discriminant algebra.

This example is familiar in the case where the characteristic of k_0 is not 2. Given a separable polynomial $f \in k[x]$, one can consider the étale k-algebra $K := k[x]/(f)$. The discriminant algebra of K—here, $\underline{\mathrm{sgn}}(K)$—is $k[x]/(x^2 - d)$, where d is the usual elementary notion of discriminant of f, i.e., the product of squares of differences of roots of f.

(For a discussion in characteristic 2, see [**Wa**].)

One of the main results of S is that $\mathscr{S}_n$ "only has mod 2 invariants", see S24.12.

(2) Let G be a semisimple algebraic group over k_0. It fits into an exact sequence

$$1 \longrightarrow C \longrightarrow \widetilde{G} \longrightarrow G \longrightarrow 1,$$

[a] We systematically refer to specific contents of S by S followed by a reference number. For example, Proposition 16.2 on page 39 will be referred to as S16.2.

[b] Below, we only consider algebraic groups that are linear.

where $\widetilde{G}$ is simply connected and C is finite and central in $\widetilde{G}$. This gives a connecting homomorphism in Galois cohomology

$$H^1(k, G) \xrightarrow{\delta} H^2(k, C)$$

that defines an invariant $\delta\colon H^1(*, G) \to H^2(*, C)$.

(3) The map e_n that sends the n-Pfister quadratic form $\langle 1, -a_1\rangle \otimes \cdots \otimes \langle 1, -a_n\rangle$ (over a field k of characteristic $\neq 2$) to the class $(a_1)\cdot(a_2)\cdots\cdot(a_n)$ in $H^n(k, \mathbb{Z}/2\mathbb{Z})$ depends only on the isomorphism class of the quadratic form. (Compare §18 of S.)

The Milnor Conjecture (now a theorem, see [**Vo**, 7.5] and [**OVV**, 4.1]) states that e_n extends to a well-defined additive map

$$e_n\colon I^n \to H^n(k, \mathbb{Z}/2\mathbb{Z})$$

that is zero on I^{n+1} and induces an isomorphism $I^n/I^{n+1} \xrightarrow{\sim} H^n(k, \mathbb{Z}/2\mathbb{Z})$. (Here I^n denotes the n-th power of the ideal I of even-dimensional forms in the Witt ring of k.)

(4) For G a quasi-simple simply connected algebraic group, there is an invariant $r_G : H^1(*, G) \to H^3(*, \mathbb{Q}/\mathbb{Z}(2))$ called the *Rost invariant*. It is the main subject of [**Mer**]. When G is Spin_n, i.e., a split simply connected group of type B or D, the Rost invariant amounts to the invariant e_3 in (3) above, cf. [**Mer**, 2.3].

The Rost invariant has the following useful property: If G' is also a quasi-simple simply connected algebraic group and $\rho : G' \to G$ is a homomorphism, then the composition

$$H^1(*, G') \xrightarrow{\rho} H^1(*, G) \xrightarrow{r_G} H^3(*, \mathbb{Q}/\mathbb{Z}(2))$$

equals $n_\rho r_{G'}$ for some natural number n_ρ, called the *Rost multiplier* of ρ, see [**Mer**, p. 122]. (The number n_ρ was defined in [**Dy 57b**, §2] and is often called the "Dynkin index" of ρ. We call it the Rost multiplier following [**Mer**] because that name describes what n_ρ does: it multiplies the Rost invariant.)

(5) Suppose that k_0 contains a primitive 4-th root of unity. The trace quadratic form on a central simple algebra A of degree[c] 4 is Witt-equivalent to a direct sum $q_2 \oplus q_4$ where q_i is an i-Pfister form, see [**RST**]. The maps $f_i : A \mapsto e_i(q_i)$ define invariants $H^1(*, \mathrm{PGL}_4) \to H^i(*, \mathbb{Z}/2\mathbb{Z})$ for $i = 2$ and 4. Rost-Serre-Tignol prove that $f_2(A)$ is zero if and only if $A \otimes A$ is a matrix algebra and $f_4(A)$ is zero if and only if A is cyclic.[d]

(For the case where k_0 has characteristic 2, see [**Tig**].)

1.3. Let C be a finite $\mathrm{Gal}(k_0)$-module[e] of exponent not divisible by the characteristic of k_0. We define a functor M by setting

$$M^d(k, C) := H^d(k, C(d-1))$$

---

[c] Recall that the dimension of a central simple algebra over its center is always a square integer, and the *degree* is the square root of the dimension.

[d] The term "cyclic" is defined in 6.4 below.

[e] We write $\mathrm{Gal}(k_1/k_0)$ for the group of k_0-automorphisms of an extension k_1/k_0 and $\mathrm{Gal}(k_0)$ for the absolute Galois group of k_0, i.e., the group of k_0-automorphisms of a separable closure of k_0.

where $C(d-1)$ denotes the $(d-1)$-st Tate twist of C as in S7.8 and

$$M(k,C) := \bigoplus_{d\geq 0} M^d(k,C).$$

We are mainly interested in

$$M(k,\mathbb{Z}/n\mathbb{Z}) = H^0(k,\mathrm{Hom}(\boldsymbol{\mu}_n,\mathbb{Z}/n\mathbb{Z})) \oplus H^1(k,\mathbb{Z}/n\mathbb{Z}) \oplus \bigoplus_{d\geq 2} H^d(k,\boldsymbol{\mu}_n^{\otimes(d-1)}).$$

Many invariants take values in $M(*,\mathbb{Z}/n\mathbb{Z})$, for example:

(2bis) For $G = \mathrm{PGL}_n$, the invariant δ in Example 1.2.2 is

$$\delta\colon H^1(*,\mathrm{PGL}_n) \to H^2(*,\boldsymbol{\mu}_n) \subset M(*,\mathbb{Z}/n\mathbb{Z}).$$

We remark that $H^2(k,\boldsymbol{\mu}_n)$ can be identified with the n-torsion in the Brauer group of k via Kummer theory.

(4bis) Let G be a group as in 1.2.4. Write i for the Dynkin index of G as in [**Mer**, p. 130] and put $n := i$ if $\operatorname{char} k_0 = 0$ or $i = p^\ell n$ for n not divisible by p if $\operatorname{char} k_0$ is a prime p. The Rost invariant maps

$$H^1(*,G) \to H^3(*,\boldsymbol{\mu}_n^{\otimes 2}) \subset M(*,\mathbb{Z}/n\mathbb{Z}).$$

(6) If the characteristic of k_0 is different from 2, then $\mathbb{Z}/2\mathbb{Z}(d)$ equals $\mathbb{Z}/2\mathbb{Z}$ for every d, and $M(k,\mathbb{Z}/2\mathbb{Z})$ is the mod 2 cohomology ring $H^\bullet(k,\mathbb{Z}/2\mathbb{Z})$. Most of the cohomological invariants considered in S take values in the functor $k \mapsto M(k,\mathbb{Z}/2\mathbb{Z})$.

We remark that for n dividing 24, $\boldsymbol{\mu}_n^{\otimes 2}$ is isomorphic to $\mathbb{Z}/n\mathbb{Z}$ [**KMRT**, p. 444, Ex. 11]. In that case,

$$M^d(k,\mathbb{Z}/n\mathbb{Z}) \cong \begin{cases} H^0(k,\mathrm{Hom}(\boldsymbol{\mu}_n,\mathbb{Z}/n\mathbb{Z})) & \text{if } d=0 \\ H^d(k,\mathbb{Z}/n\mathbb{Z}) & \text{if } d \text{ is odd} \\ H^d(k,\boldsymbol{\mu}_n) & \text{if } d \text{ is even and } d\neq 0 \end{cases}$$

The other main target for invariants is $M(*,\boldsymbol{\mu}_n)$, characterized by

$$M(k,\boldsymbol{\mu}_n) = \mathbb{Z}/n\mathbb{Z} \oplus \bigoplus_{d\geq 1} H^d(k,\boldsymbol{\mu}_n^{\otimes d}).$$

This is naturally a ring, and we write $R_n(k)$ for $M(k,\boldsymbol{\mu}_n)$ when we wish to view it as such. (This ring is a familiar one: the Bloch-Kato Conjecture asserts that it is isomorphic to the quotient $K_\bullet^M(k)/n$ of the Milnor K-theory ring $K_\bullet^M(k)$.) When C is n-torsion, the abelian group $M(k,C)$ is naturally an $R_n(k)$-module.

For various algebraic groups G and Galois-modules C, we will determine the invariants $H^1(*,G) \to M(*,C)$. We abuse language by calling these "invariants of G with values in C", "C-invariants of G", etc. We write $\mathrm{Inv}(G,C)$ or $\mathrm{Inv}_{k_0}(G,C)$ for the collection of such invariants.[f] For example, the invariants in (2bis) and (4bis) above belong to $\mathrm{Inv}(\mathrm{PGL}_n,\mathbb{Z}/n\mathbb{Z})$ and $\mathrm{Inv}(G,\mathbb{Z}/n\mathbb{Z})$ respectively. Note that $\mathrm{Inv}(G,C)$ is an abelian group for every algebraic group G, and, when C is n-torsion, $\mathrm{Inv}_{k_0}(G,C)$ is an $R_n(k_0)$-module.

1.4. Constant and normalized. Fix an element $m \in M(k_0,C)$. For every group G, the collection of maps that sends every element of $H^1(k,G)$ to the image of

[f]Strictly speaking, this notation disagrees with the notation defined on page 11 of S. But there is no essential difference, because in S the target C is nearly always taken to be $\mathbb{Z}/2\mathbb{Z}$, and $\mathbb{Z}/2\mathbb{Z}(d)$ is canonically isomorphic to $\mathbb{Z}/2\mathbb{Z}$ for all d.

m in $M(k, C)$ for every extension k/k_0 is an invariant in $\mathrm{Inv}(G, C)$. Such invariants are called *constant*. The constant invariant given by the element zero in $M(k_0, C)$ is called the *zero invariant*; it is the identity element of the group $\mathrm{Inv}(G, C)$.

An invariant $a \in \mathrm{Inv}(G, C)$ is *normalized* if a sends the neutral class in $H^1(k, G)$ to zero in $M(k, C)$ for every extension k/k_0. We write $\mathrm{Inv}^{\mathrm{norm}}(G, C)$ for the normalized invariants in $\mathrm{Inv}(G, C)$.

The reader can find a typical application of cohomological invariants in Appendix A.

2. Invariants of $\boldsymbol{\mu}_n$

Fix a natural number n not divisible by the characteristic of the field k_0. In this section, we determine the invariants of $\boldsymbol{\mu}_n$ with values in $\boldsymbol{\mu}_n$, along with a small variation.

There are two obvious invariants of $\boldsymbol{\mu}_n$:

(1) The constant invariant (as in 1.4) that sends everything in $H^1(k, \boldsymbol{\mu}_n)$ to the element 1 in $\mathbb{Z}/n\mathbb{Z} \subset M(k, \boldsymbol{\mu}_n)$.

(2) The invariant $\underline{\mathrm{id}}$ that is the identity map

$$H^1(k, \boldsymbol{\mu}_n) \to H^1(k, \boldsymbol{\mu}_n) \subset M(k, \boldsymbol{\mu}_n)$$

for every k/k_0.

2.1. PROPOSITION. *$\mathrm{Inv}_{k_0}(\boldsymbol{\mu}_n, \boldsymbol{\mu}_n)$ is a free $R_n(k_0)$-module with basis $1, \underline{\mathrm{id}}$.*

This proposition can easily be proved by adapting the proof of S16.2. Alternatively, it is [**MPT**, Cor. 1.2]. In the interest of exposition, we give an elementary proof in the case where k_0 is algebraically closed.

For every extension K/k_0, the group $H^1(K, \boldsymbol{\mu}_n)$ is identified with $K^\times/K^{\times n}$ by Kummer theory. For an element $x \in K^\times$, we write (x) for the corresponding element of $H^1(K, \boldsymbol{\mu}_n)$; the field K and the number n can be inferred from context.

We need the following lemma, which is a special case of S12.3.

2.2. LEMMA. *If invariants $a, a' \in \mathrm{Inv}_{k_0}(\boldsymbol{\mu}_n, \boldsymbol{\mu}_n)$ agree on $(t) \in H^1(k_0(t), \boldsymbol{\mu}_n)$, then a and a' are equal.*

PROOF. Replacing a, a' with $a - a', 0$ respectively, we may assume that a' is the zero invariant.

Fix an extension E of k_0 and an element $y \in E^\times$. Write M for the functor $M(*, \boldsymbol{\mu}_n)$ as in 1.3 and consider the commutative diagram

$$\begin{array}{ccccc} H^1(E, \boldsymbol{\mu}_n) & \longrightarrow & H^1(E((t-y)), \boldsymbol{\mu}_n) & \longleftarrow & H^1(k_0(t), \boldsymbol{\mu}_n) \\ \big\downarrow a_E & & \big\downarrow a_{E((t-y))} & & \big\downarrow a_{k_0(t)} \\ M(E) & \longrightarrow & M(E((t-y))) & \longleftarrow & M(k_0(t)) \end{array} \tag{2.3}$$

The polynomial $x^n - y/t$ in $E((t-y))[x]$ has residue $x^n - 1$ in $E[x]$, which has a simple root, namely $x = 1$. Therefore $x^n - y/t$ has a root over $E((t-y))$ by Hensel's Lemma, and the images of $(y) \in H^1(E, \boldsymbol{\mu}_n)$ and $(t) \in H^1(k_0(t), \boldsymbol{\mu}_n)$ in $H^1(E((t-y)), \boldsymbol{\mu}_n)$ agree. The commutativity of the diagram implies that the image of (y) in $M(E((t-y)))$ is the same as the image of $a_{k_0(t)}(t)$, i.e., zero. But the map $M(E) \to M(E((t-y)))$ is an injection by S7.7, so $a_E(y)$ is zero. This proves the lemma. □

PROOF OF PROP. 2.1. We assume that k_0 is algebraically closed. Fix an invariant $a \in \mathrm{Inv}_{k_0}(\boldsymbol{\mu}_n, \boldsymbol{\mu}_n)$, and consider the torsor class $(t) \in H^1(k_0(t), \boldsymbol{\mu}_n)$. We claim that $a(t)$ is unramified away from $\{0, \infty\}$. Indeed, any other point on the affine line over k_0 is an ideal $(t - y)$ for some $y \in k_0^\times$ because k_0 is algebraically closed. Consider the diagram (2.3) with the E's replaced with k_0's. As in the proof of Lemma 2.2, the images of $(y) \in H^1(k_0, \boldsymbol{\mu}_n)$ and $(t) \in H^1(k_0(t), \boldsymbol{\mu}_n))$ agree in $H^1(k_0((t-y)), \boldsymbol{\mu}_n)$ by Hensel's Lemma, hence the image of (t) in $M(k_0((t-y)))$ comes from $M(k_0)$. That is, $a(t)$ is unramified at $(t-y)$. This proves the claim, and by S9.4 we have:

$$a(t) = \lambda_0 + \lambda_1 \cdot (t)$$

for uniquely determined elements $\lambda_0, \lambda_1 \in M(k_0)$.

Put $a' := \lambda_0 \cdot 1 + \lambda_1 \cdot \underline{\mathrm{id}}$. Since the invariants a, a' agree on (t), the two invariants are the same by Lemma 2.2. This proves that $1, \underline{\mathrm{id}}$ span $\mathrm{Inv}_{k_0}(\boldsymbol{\mu}_n, \boldsymbol{\mu}_n)$.

As for linear independence, suppose that the invariant $\lambda_0 \cdot 1 + \lambda_1 \cdot \underline{\mathrm{id}}$ is zero. Then λ_0—the value of a on the trivial class—is zero. The other coefficient, λ_1, is the residue at $t = 0$ of $a(t)$ in $M(k_0)$. □

Recall from S4.5 that every invariant can be written uniquely as (constant) + (normalized). Clearly, the proposition shows that $\mathrm{Inv}_{k_0}^{\mathrm{norm}}(\boldsymbol{\mu}_n, \boldsymbol{\mu}_n)$ is a free $R_n(k_0)$-module with basis $\underline{\mathrm{id}}$.

Really, the proof of Prop. 2.1 given above is the same as the proof of S16.2 in the case where k_0 is algebraically closed, except that we have unpacked the references to S11.7 and S12.3 (which are both elaborations of the Rost Compatibility Theorem) with the core of the Rost Compatibility Theorem that is sufficient in this special case.

2.4. REMARK. The argument using Hensel's Lemma in the proof of Lemma 2.2 has real problems when the characteristic of k_0 divides n. For example, when the characteristic of k_0 (and hence E) is a prime p, the element t/y has no p-th root in $E((t-y))$ for every $y \in E^\times$. Speaking very roughly, this is the reason for the global assumption that the characteristic of k_0 does not divide the exponent of C.

2.5. $\boldsymbol{\mu}_n$ INVARIANTS OF $\boldsymbol{\mu}_{sn}$. Let s be a positive integer not divisible by the characteristic of k_0. The s-th power map (the natural surjection) $s \colon \boldsymbol{\mu}_{sn} \to \boldsymbol{\mu}_n$ fits into a commutative diagram

$$\begin{array}{ccccccccc} 1 & \longrightarrow & \boldsymbol{\mu}_{sn} & \longrightarrow & \mathbb{G}_m & \xrightarrow{sn} & \mathbb{G}_m & \longrightarrow & 1 \\ & & {\scriptstyle s}\downarrow & & {\scriptstyle s}\downarrow & & \| & & \\ 1 & \longrightarrow & \boldsymbol{\mu}_n & \longrightarrow & \mathbb{G}_m & \xrightarrow{n} & \mathbb{G}_m & \longrightarrow & 1 \end{array} \tag{2.6}$$

It induces an invariant $\underline{s} \colon H^1(*, \boldsymbol{\mu}_{sn}) \to H^1(*, \boldsymbol{\mu}_n)$. A diagram chase on (2.6) shows that for each k, $\underline{s}$ is the surjection

$$k^\times / k^{\times sn} \to k^\times / k^{\times n} \quad \text{given by } x k^{\times sn} \mapsto x k^{\times n}.$$

The proof of Prop. 2.1 with obvious modifications gives:

PROPOSITION. *$\mathrm{Inv}_{k_0}(\boldsymbol{\mu}_{sn}, \boldsymbol{\mu}_n)$ is a free $R_n(k_0)$-module with basis $1, \underline{s}$.* □

We will apply this in 18.7 and §20 below in the case $s = n = 2$. In that case, we will continue to write $\underline{s}$ instead of the more logical $\underline{2}$.

2.7. EXERCISE. Let C be a finite $\mathrm{Gal}(k_0)$-module whose order is a power of n. For $x \in M(k_0, C)$ such that $nx = 0$, define a cup product

$$- \bullet x \colon H^1(k, \boldsymbol{\mu}_n) \to M(k, C)$$

by mimicking §23 of S. This defines an invariant of $\boldsymbol{\mu}_n$ that we denote by $\underline{\mathrm{id}} \bullet x$. Prove that every normalized invariant $H^1(*, \boldsymbol{\mu}_n) \to M(*, C)$ is of the form $\underline{\mathrm{id}} \bullet x$ for a unique $x \in M(k_0, C)$ with $nx = 0$.

3. Quasi-Galois extensions and invariants of $\mathbb{Z}/p\mathbb{Z}$

Here and for the rest of Part I, we assume that *C is a finite $\mathrm{Gal}(k_0)$-module of exponent not divisible by the characteristic of k_0.*

3.1. Let $p_1, p_2, \ldots, p_r$ be the distinct primes dividing the exponent of C. There is a canonical identification $C = \prod_{i=1}^r {}_{p_i}C$, where ${}_{p_i}C$ denotes the submodule of C consisting of elements of order a power of p_i. This gives an identification

$$\mathrm{Inv}_{k_0}(G, C) = \prod_{i=1}^{r} \mathrm{Inv}_{k_0}(G, {}_{p_i}C)$$

that is functorial with respect to changes in the field k_0 and the group G.

3.2. LEMMA. *If k_1 is a finite extension of k_0 of dimension relatively prime to the exponent of C, then the natural map*

$$\mathrm{Inv}_{k_0}(G, C) \to \mathrm{Inv}_{k_1}(G, C)$$

is an injection.

PROOF. By 3.1, we may assume that the exponent of C is a power of a prime p.

Let a be an invariant in the kernel of the displayed map. Fix an extension E/k_0 and an element $x \in H^1(E, G)$; we show that $a(x)$ is zero in $M(E, C)$, hence a is the zero invariant.

First suppose that k_1/k_0 is separable. The tensor product $E \otimes_{k_0} k_1$ is a direct product of fields $E_1 \times E_2 \times \cdots \times E_r$ (since k_1 is separable over k_0), and at least one of them—say, E_i—has dimension over E not divisible by p (because p does not divide $[k_1 : k_0]$). We have

$$\mathrm{res}_{E_i/E}\, a(x) = a(\mathrm{res}_{E_i/E}\, x) = 0$$

because k_1 injects into E_i. But the dimension $[E_i : E]$ is not divisible by p, so $a(x)$ is zero in $M(E, C)$.

If k_1/k_0 is purely inseparable, then there is a compositum E_1 of E and k_1 such that E_1/E is purely inseparable. As p is different from the characteristic of k_0, the restriction $M(E, C) \to M(E_1, C)$ is injective, and again we conclude that $a(x)$ is zero in $M(E, C)$.

In the general case, let k_s be the separable closure of k_0 in k_1. The map displayed in the lemma is the composition

$$\mathrm{Inv}_{k_0}(G, C) \to \mathrm{Inv}_{k_s}(G, C) \to \mathrm{Inv}_{k_1}(G, C),$$

and both arrows are injective by the preceding two paragraphs. Hence the composition is injective. □

3.3. Suppose that k_1/k_0 is finite of dimension relatively prime to the exponent of C as in 3.2, and suppose further that k_1/k_0 is quasi-Galois (= normal), i.e., k_1 is the splitting field for a collection of polynomials in $k_0[x]$. The separable closure k_s of k_0 in k_1 is a Galois extension of k_0. (See [**Bou Alg**, §V.11, Prop. 13] for the general structure of k_1/k_0.) We write $\mathrm{Gal}(k_1/k_0)$ for the group of k_0-automorphisms of k_1.

The group $\mathrm{Gal}(k_1/k_0)$ acts on $H^1(k_1, G)$ as follows. An element $g \in \mathrm{Gal}(k_1/k_0)$ sends a 1-cocycle b to a 1-cocycle $g * b$ defined by

$$(g * b)_s = {}^g b_{g^{-1}s}.$$

The Galois group acts similarly acts on $M(k_1, C)$, see e.g. [**We**, Cor. 2-3-3].

LEMMA. *If k_1/k_0 is finite quasi-Galois and $[k_1 : k_0]$ is relatively prime to the exponent of C, then the restriction map*

$$M(k_0, C) \to M(k_1, C)$$

identifies $M(k_0, C)$ with the subgroup of $M(k_1, C)$ consisting of elements fixed by $\mathrm{Gal}(k_1/k_0)$. □

PROOF. Write k_i for the maximal purely inseparable subextension of k_1/k_0; the extension k_1/k_i is Galois. It is standard that the restriction map $M(k_i, C) \to M(k_1, C)$ identifies $M(k_i, C)$ with the $\mathrm{Gal}(k_1/k_i)$-fixed elements of $M(k_1, C)$. To complete the proof, it suffices to note that restriction identifies $\mathrm{Gal}(k_1/k_i)$ with $\mathrm{Gal}(k_1/k_0)$ and $M(k_0, C)$ with $M(k_i, C)$, because k_i/k_0 is purely inseparable. □

3.4. INVARIANTS UNDER QUASI-GALOIS EXTENSIONS. Continue the assumption that k_1 is a finite quasi-Galois extension of k_0. For every extension E of k_0, there is—up to k_0-isomorphism—a unique compositum E_1 of E and k_1; the field E_1 is quasi-Galois over E and $\mathrm{Gal}(E_1/E)$ is identified with a subgroup of $\mathrm{Gal}(k_1/k_0)$. We say that an invariant $a \in \mathrm{Inv}_{k_1}(G, C)$ is *Galois-fixed* if for every E/k_0, $x \in H^1(E_1, G)$, and $g \in \mathrm{Gal}(E_1/E)$, we have

$$g * a(g^{-1} * x) = a(x) \quad \in M(E_1, C).$$

PROPOSITION. *If k_1/k_0 is finite quasi-Galois and $[k_1 : k_0]$ is relatively prime to the exponent of C, then the restriction map*

$$\mathrm{Inv}_{k_0}(G, C) \to \mathrm{Inv}_{k_1}(G, C)$$

identifies $\mathrm{Inv}_{k_0}(G, C)$ with the subgroup of Galois-fixed invariants in $\mathrm{Inv}_{k_1}(G, C)$.

PROOF. The restriction map is an injection by Lemma 3.2.

Fix an invariant $a_1 \in \mathrm{Inv}_{k_1}(G, C)$. If a_1 is the restriction of an invariant defined over k_0, then a commutes with every morphism in $\mathrm{Aut}_{\mathsf{Fields}/E}(E_1)$ for every extension E/k_0, i.e., a_1 is Galois-fixed.

To prove the converse, suppose that a_1 is Galois-fixed. For $x \in H^1(E, G)$ and $g \in \mathrm{Gal}(E_1/E)$, we have

$$g * a_1(\mathrm{res}_{E_1/E}\, x) = a_1(g * \mathrm{res}_{E_1/E}\, x) = a_1(\mathrm{res}_{E_1/E}\, x) \quad \in M(E_1, C)$$

since a_1 is Galois-fixed. Lemma 3.3 gives that $a_1(\mathrm{res}_{E_1/E}\, x)$ is the restriction of a unique element $a_0(x)$ in $M(E, C)$. In this way, we obtain a function $H^1(E, G) \to M(E, C)$. It is an exercise to verify that this defines an invariant $a_0\colon H^1(*, G) \to M(*, C)$. Clearly, the restriction of a_0 to k_1 is a_1. □

3.5. Continue the assumption that k_1/k_0 is finite quasi-Galois and $[k_1 : k_0]$ is relatively prime to the exponent of C.

We fix a natural number n not divisible by the characteristic of k_0 such that $nC = 0$, and we suppose that $\mathrm{Inv}_{k_0}^{\mathrm{norm}}(G, C)$ contains elements $a_1, a_2, \ldots, a_r$ whose restrictions form an $R_n(k_1)$-basis of $\mathrm{Inv}_{k_1}^{\mathrm{norm}}(G, C)$. We find:

COROLLARY. *$a_1, a_2, \ldots, a_r$ is an $R_n(k_0)$-basis of* $\mathrm{Inv}_{k_0}^{\mathrm{norm}}(G, C)$.

[Clearly, the corollary also holds if one can replaces $\mathrm{Inv}^{\mathrm{norm}}$ with Inv throughout.]

PROOF. Since k_1 is finite quasi-Galois over k_0, restriction identifies $R_n(k_0)$ with the $\mathrm{Gal}(k_1/k_0)$-fixed elements in $R_n(k_1)$ (by Lemma 3.3 with $C = \boldsymbol{\mu}_n$) and the natural map

$$\mathrm{Inv}_{k_0}(G, C) \to \mathrm{Inv}_{k_1}(G, C) \tag{3.6}$$

is an injection by Prop. 3.4.

Let $\lambda_1, \lambda_2, \ldots, \lambda_r \in R_n(k_0)$ be such that $\sum \lambda_i a_i$ is zero in $\mathrm{Inv}_{k_0}(G, C)$. Every λ_i is killed by k_1, hence λ_i is zero in $R_n(k_0)$ for all i. This proves that the a_i are linearly independent over k_0.

As for spanning, let a be in $\mathrm{Inv}_{k_0}(G, C)$. The restriction of a to k_1 equals $\sum \lambda_i a_i$ for some $\lambda_i \in R_n(k_1)$. But a is fixed by $\mathrm{Gal}(k_1/k_0)$, hence so are the λ_i, i.e., λ_i is the restriction of an element of $R_n(k_0)$ which we may as well denote also by λ_i. Since $a - \sum \lambda_i a_i$ is zero over k_1, it is zero over k_0. This proves that the a_i span over k_0. □

3.7. PROPOSITION. *If p is a prime not equal to the characteristic of k, then $\mathrm{Inv}_{k_0}^{\mathrm{norm}}(\mathbb{Z}/p\mathbb{Z}, \mathbb{Z}/p\mathbb{Z})$ is a free $R_p(k_0)$-module with basis $\underline{\mathrm{id}}$.*

[The reader may wonder why we have switched to describing the normalized invariants, whereas in the proposition above and in S, the full module of invariants was described. The difficulty is that here the invariants are taking values in

$$H^0(*, \mathrm{Hom}(\boldsymbol{\mu}_p, \mathbb{Z}/p\mathbb{Z})) \oplus H^1(*, \mathbb{Z}/p\mathbb{Z}) \oplus H^2(*, \boldsymbol{\mu}_p) \oplus \cdots,$$

and it is not clear how to specify a basis for the constant invariants.]

PROOF. If k_0 contains a primitive p-th root of unity, then we may use it to identify $\mathbb{Z}/p\mathbb{Z}$ with $\boldsymbol{\mu}_p$ and apply Prop. 2.1.

For the general case, take k_1 to be the extension obtained by adjoining a primitive p-th root of unity; it is a Galois extension of dimension not divisible by p, and the proposition holds for k_1 by the previous paragraph. Cor. 3.5 finishes the proof. □

3.8. EXERCISE. Extend Prop. 3.7 by describing $\mathrm{Inv}_{k_0}^{\mathrm{norm}}(\mathbb{Z}/n\mathbb{Z}, \mathbb{Z}/n\mathbb{Z})$, where n is square-free and not divisible by the characteristic.

3.9. EXERCISE. Let k_0 be a field of characteristic zero. What are the mod 2 invariants of the dihedral group G of order 8? That is, what is $\mathrm{Inv}_{k_0}^{\mathrm{norm}}(G, \mathbb{Z}/2\mathbb{Z})$?

[Note that G is the Weyl group of a root system of type B_2, so one may apply S25.15: an invariant of G is determined by its restriction to the elementary abelian 2-subgroups of G.]

4. An example: the mod p Bockstein map

Let p be a prime not equal to the characteristic of k. (We allow $p = 2$ or $\operatorname{char} k = 2$.) Consider the sequence of $\operatorname{Gal}(k)$-modules (with trivial action):

$$(4.1) \qquad 0 \longrightarrow \mathbb{Z}/p\mathbb{Z} \longrightarrow \mathbb{Z}/p^2\mathbb{Z} \longrightarrow \mathbb{Z}/p\mathbb{Z} \longrightarrow 0.$$

Let d be the coboundary (a.k.a. the "Bockstein map") defined by this sequence; it maps $H^n(k, \mathbb{Z}/p\mathbb{Z})$ into $H^{n+1}(k, \mathbb{Z}/p\mathbb{Z})$. For $n = 1$, we may view it as a cohomological invariant on $H^1(k, \mathbb{Z}/p\mathbb{Z})$ with values in $H^2(k, \mathbb{Z}/p\mathbb{Z})$. By the same proof as Prop. 3.7, there exists an element $c \in H^1(k_0, \mathbb{Z}/p\mathbb{Z})$ (where k_0 is the prime field) such that $d(x) = x \cdot c$ for every extension k/k_0 and every $x \in H^1(k, \mathbb{Z}/p\mathbb{Z})$.

A natural question is: *what is c?* In the case $p = 2$, the answer is well-known: $c = (-1)$, the class of -1 in $H^1(k_0, \mathbb{Z}/2\mathbb{Z}) = k_0^\times/k_0^{\times 2}$. We give the answer in general.

4.2. Let $U = \mathbb{Z}_p^\times$ be the group of p-adic units; there is a unique homomorphism $e \colon U \to \mathbb{Z}/p\mathbb{Z}$ such that

$$e(1+p) = 1 \quad \text{and} \quad e(1+p^2) = 0.$$

We may write it explicitly as:

$$e(u) = (1 - u^{p-1})/p \pmod{p}.$$

Write χ for the cyclotomic character $\operatorname{Gal}(k_0) \to U$. The composite map $c = e \circ \chi$ belongs to $\operatorname{Hom}(\operatorname{Gal}(k_0), \mathbb{Z}/p\mathbb{Z}) = H^1(k_0, \mathbb{Z}/p\mathbb{Z})$. This is the element c we were looking for. Namely:

THEOREM. *$d(x) = x \cdot e\chi$ for every extension k/k_0 and every x in $H^1(k, \mathbb{Z}/p\mathbb{Z})$.*

We prove the theorem at the end of this section by comparing d with the coboundary map

$$d^* : H^1(k, \boldsymbol{\mu}_p) \to H^2(k, \boldsymbol{\mu}_p).$$

arising from the exact sequence

$$(4.3) \qquad 1 \longrightarrow \boldsymbol{\mu}_p \longrightarrow \boldsymbol{\mu}_{p^2} \longrightarrow \boldsymbol{\mu}_p \longrightarrow 1.$$

Note that, if we identify $\boldsymbol{\mu}_p$ with $\mathbb{Z}/p\mathbb{Z}$, both d and d^* map $H^1(k, \mathbb{Z}/p\mathbb{Z})$ into $H^2(k, \mathbb{Z}/p\mathbb{Z})$. The comparison of d and d^* can be done in a more general setting, as follows:

4.4. Consider a group G and an exact sequence of G-modules

$$(4.5) \qquad 0 \longrightarrow A \longrightarrow B \xrightarrow{\ q\ } C \longrightarrow 0,$$

together with a 1-cocycle $z \colon G \to \operatorname{Hom}(C, A)$. We may use z to twist the action of G on B: if $s \in G$ and $b \in B$, the "new" transform of b by s, which we shall denote by $s \circ b$, is

$$s \circ b = sb + z(s)(s.q(b)).$$

One checks that this defines an action of G on B. Call B_z the G-module so defined. We have an exact sequence

$$(4.6) \qquad 0 \longrightarrow A \longrightarrow B_z \longrightarrow C \xrightarrow{\ q\ } 0,$$

where the maps $A \to B_z$ and $B_z \to C$ are the same as before. (Those maps are compatible with the new G-action on B_z.)

The exact sequences (4.5) and (4.6) define coboundary maps d and d_z mapping $H^n(G, C)$ into $H^{n+1}(G, A)$.

LEMMA. *We have*

$$d_z(x) = d(x) + z\cdot x$$

for every $x \in H^n(G, C)$.

(The cup product $z\cdot x$ is relative to the natural pairing $\mathrm{Hom}(C, A) \times C \to A$.)

PROOF. This is a simple cochain computation. One chooses an n-cocycle f_C representing x and lifts it to an n-cochain f with values in B. The cohomology class $d(x)$ is represented by the $(n+1)$-cocycle

$$(s_1, \ldots) \mapsto s_1.f(s_2, \ldots) - f(s_1 s_2, \ldots) + f(s_1, \ldots);$$

similarly, $d_z(x)$ is represented by

$$(s_1, \ldots) \mapsto s_1 \circ f(s_2, \ldots) - f(s_1 s_2, \ldots) + f(s_1, \ldots).$$

These expressions differ only by their first terms (which involve respectively an s_1-transform and a twisted s_1-transform). Hence $d_z(x) - d(x)$ is represented by the $(n+1)$-cocycle

$$(s_1, \ldots) \mapsto s_1 \circ f(s_2, \ldots) - s_1.f(s_2, \ldots) = z(s_1)(s_1.f_C(s_2, \ldots)),$$

which is the cup product of the 1-cocycle z and the n-cocycle f_C. □

PROOF OF THEOREM 4.2. By Lemma 3.2, we may assume that k contains a primitive p-th root of unity ζ. Use ζ to identify $\mathbb{Z}/p\mathbb{Z}$ with $\boldsymbol{\mu}_p$.

Let ω denote a p-th root of ζ in our given separable closure of k. For $s \in \mathrm{Gal}(k)$, the element $s(\omega)\omega^{-1}$ is a p-th root of unity, hence $s(\omega) = \omega^{pr+1}$ for some $0 \leq r < p$. The map $\omega \mapsto 1$ defines an isomorphism of $\boldsymbol{\mu}_{p^2}$ with $(\mathbb{Z}/p^2\mathbb{Z})_z$ — with $\mathbb{Z}/p^2\mathbb{Z}$ twisted as in 4.4 — for z defined by $z(s)(1) = r$. The equality $\mathrm{Hom}(\mathbb{Z}/p\mathbb{Z}, \mathbb{Z}/p\mathbb{Z}) = \mathbb{Z}/p\mathbb{Z}$ identifies z with $e\chi$ and Lemma 4.4 gives that

$$d(x) = d^*(x) - e\chi\cdot x.$$

However, the natural map $H^1(k, \boldsymbol{\mu}_{p^2}) \to H^1(k, \boldsymbol{\mu}_p)$ is surjective, so $d^*(x)$ is zero. The theorem follows by the skew-commutativity of the cup-product. □

4.7. REMARK. One can also give a general formula for the Bockstein map in any dimension. More precisely, say that an element x of $H^n(k, \mathbb{Z}/p\mathbb{Z})$ is *almost decomposable* if there is an extension k'/k of dimension prime to p such that the image of x in $H^n(k', \mathbb{Z}/p\mathbb{Z})$ is a sum of decomposable elements (products of n elements of degree 1). (It follows from a conjecture of Bloch-Kato that every element of $H^n(k, \mathbb{Z}/p\mathbb{Z})$ is almost decomposable.) We have

$$d(x) = (-1)^{n-1} n\, x\cdot e\chi$$

for every x in $H^n(k, \mathbb{Z}/p\mathbb{Z})$ that is almost decomposable. This is proved by induction on n, using the derivation property of the Bockstein map [**Lang**, pp. 83–85], namely:

$$d(x_1\cdot x_2) = d(x_1)\cdot x_2 + (-1)^{\deg(x_1)} x_1\cdot d(x_2).$$

Note that when p divides n, we have $d(x) = 0$ provided that x is almost decomposable.

4.8. EXAMPLE. When k contains a primitive p-th root of unity ζ, one can use ζ to identify $\mathbb{Z}/p\mathbb{Z}$ with $\boldsymbol{\mu}_p$. Then the class $e\chi$ in $H^1(k, \mathbb{Z}/p\mathbb{Z}) = H^1(k, \boldsymbol{\mu}_p)$ can be identified with an element of $k^\times/k^{\times p}$. Which element? In the notation of the proof of Theorem 4.2, we have:

$$s(\omega) = \zeta^{z(s)}\omega.$$

That is, $e\chi$ is identified with the class of ζ.

5. Restricting invariants

Let A and A' be functors $\mathsf{Fields}_{/k_0} \to \mathsf{Sets}$, and fix a morphism $\phi\colon A' \to A$. For example, a homomorphism of algebraic groups $G' \to G$ induces such a morphism of functors $H^1(*, G') \to H^1(*, G)$.

5.1. DEFINITION. The map ϕ is *surjective at p* (for p a prime) if for every extension k_1/k_0 and every $x \in A(k_1)$ there is a finite extension k_2 of k_1 such that

(1) $\mathrm{res}_{k_2/k_1}(x) \in A(k_2)$ is $\phi(x')$ for some $x' \in A'(k_2)$ and
(2) the dimension $[k_2 : k_1]$ is not divisible by p.

We are interested in the following condition:

(5.2) ϕ is surjective at every prime dividing the exponent of C.

Write $\mathrm{Inv}(A, C)$ for the group of invariants $A(*) \to M(*, C)$, and similarly for A'. We have:

5.3. LEMMA. *If* (5.2) *holds, then the restriction map*

$$\phi^*\colon \mathrm{Inv}_{k_0}(A, C) \to \mathrm{Inv}_{k_0}(A', C)$$

induced by ϕ is an injection.

We will strengthen this result in Section 7.

PROOF. By 3.1, we may assume that the exponent of C is a power of a prime p and that ϕ is surjective at p. The map ϕ^* is a group homomorphism, so it suffices to prove that the kernel of ϕ^* is zero; let a be in the kernel of ϕ^*. Fix an extension k_1 of k_0 and a class $x \in A(k_1)$, and let k_2 be as in 5.1. By the assumption on a, the class $a(x) \in M(k_1, C)$ is killed by k_2. But the map $M(k_1, C) \to M(k_2, C)$ is injective by 5.1.2, so $a(x)$ is zero in $M(k_1, C)$. That is, a is the zero invariant. □

5.4. KILLABLE CLASSES. Suppose that there is a natural number e such that every element of $H^1(k, G)$ is killed by an extension of k of degree dividing e for every extension k of k_0. This happens, for example, when:

(1) $G = \mathrm{PGL}_e$, a standard result from the theory of central simple algebras
(2) G is a finite constant group and $e = |G|$, because every 1-cocycle is a homomorphism $\varphi\colon \mathrm{Gal}(k_1) \to G$ and φ is killed by the extension k_2 of k_1 fixed by $\ker\varphi$. The dimension of k_2 over k_1 equals the size of the image of φ, which divides the order of G. (Compare S15.4.)

Applying Lemma 5.3 with G' the group with one element gives: *If the exponent of C is relatively prime to e, then* $\mathrm{Inv}^{\mathrm{norm}}_{k_0}(G, C)$ *is zero.*

5.5. EXAMPLE. Suppose that k_0 is algebraically closed of characteristic zero, G is a connected algebraic group, and the exponent of C is relatively prime to the order of the Weyl group of a Levi subgroup of G. *Then* $\operatorname{Inv}(G, C)$ *is zero.* Indeed, the paper [**CGR**] gives a finite constant subgroup S of G such that the exponent of C is relatively prime to $|S|$ and the map $H^1(k, S) \to H^1(k, G)$ is surjective for every extension k of k_0. (We remark that the existence of such a subgroup S answers the question implicit in the final paragraph of S22.10.) As a consequence of the surjectivity, the restriction map $\operatorname{Inv}(G, C) \to \operatorname{Inv}(S, C)$ is an injection. Hence $\operatorname{Inv}(G, C)$ is zero by 5.4.

5.6. The previous example gives a "coarse bound" in the case where G is simple. For G simple of type E_8, the order of the Weyl group is $2^{14} \cdot 3^5 \cdot 5^2 \cdot 7$. So—roughly speaking—the previous example shows there are no nonconstant cohomological invariants mod p for $p \neq 2, 3, 5, 7$. Tits [**Tits 92**] showed that every E_8-torsor is split by an extension of degree dividing $2^9 \cdot 3^3 \cdot 5$,[g] hence by 5.4 there are no nonconstant invariants mod p also for $p = 7$.

In Table 5, for each type of exceptional group G and prime p, we give a reference for classification results regarding the invariants $\operatorname{Inv}^{\text{norm}}(G, C)$ where the exponent of C is a power of p. The preceding argument shows that $\operatorname{Inv}^{\text{norm}}(G, C)$ is zero for all exceptional G and $p \neq 2, 3, 5$.

type of G	$p = 2$	$p = 3$	$p = 5$
G_2	S18.4	X	X
F_4	S22.5	Th. 8.6	X
E_6, inner, simply connected	Exercise 22.9 in S	Th. 11.9	X
E_6, inner, adjoint	Exercise 22.9 in S	?	X
E_6, outer, simply connected	?	Exercise 11.10	X
E_6, outer, adjoint	?	?	X
E_7	?	13.2, Exercise 13.3	X
E_8	?	?	Th. 15.1

TABLE 5. References for results on $\operatorname{Inv}^{\text{norm}}(G, C)$ where G is exceptional and the exponent of C is a power of a prime p

For entries marked with an X, $\operatorname{Inv}^{\text{norm}}(G, C)$ is zero. For conjectures regarding some of the question marks, see Problems 13.4 and 15.3.

5.7. EXAMPLE (Groups of square-free order). Suppose now that k_0 is algebraically closed, G is a finite constant group, and $|G|$ is square-free and not divisible by the characteristic of k_0. Then *every normalized invariant* $H^1(*, G) \to H^d(*, C)$ *is zero for* $d > 1$. Indeed, by 3.1 and 5.4, we may assume that C is a power of a prime p dividing $|G|$. For G' a p-Sylow subgroup of G, S15.4 (a more powerful version of Lemma 5.3 that is specifically for finite groups) says the restriction map

$$\operatorname{Inv}^{\text{norm}}(G, C) \to \operatorname{Inv}^{\text{norm}}(G', C)$$

is injective. As G' is isomorphic to $\boldsymbol{\mu}_p$, Exercise 2.7 says that every normalized invariant $H^1(*, \boldsymbol{\mu}_p) \to M^d(*, C)$ can be written uniquely as $\underline{\text{id}} \bullet x$ for some p-torsion element $x \in M^{d-1}(k_0, C)$. But this last set is zero because k_0 is algebraically closed.

[g]Totaro recently perfected this result by showing that every E_8-torsor is split by an extension of degree dividing $2^6 \cdot 3^2 \cdot 5$ [**To 04**] and that this bound is best possible [**To 05**].

5.8. INEFFECTIVE BOUNDS FOR ESSENTIAL DIMENSION. Recall from S5.7 that the *essential dimension* of an algebraic group G over k_0—written $\mathrm{ed}(G)$—is the minimal transcendence degree of K/k_0, where K is the field of definition of a versal G-torsor. Cohomological invariants can be used to prove lower bounds on $\mathrm{ed}(G)$: *If k_0 is algebraically closed and there is a nonzero invariant $H^1(*, G) \to H^d(*, C)$, then* $\mathrm{ed}(G) \geq d$, see S12.4.

But this bound need not be sharp, as Example 5.7 shows. Indeed, in that example we find the lower bound $\mathrm{ed}(G) \geq 1$ for G not the trivial group. But when k_0 has characteristic zero and G is neither cyclic nor dihedral of order $2 \cdot (\text{odd})$, $\mathrm{ed}(G) \geq 2$ by [**BR**, Th. 6.2].

Another example is furnished by the alternating group A_6. The bound provided by cohomological invariants is $\mathrm{ed}(A_6) \geq 2$ but the essential dimension cannot be 2 (Serre, unpublished).

6. Mod p invariants of PGL_p

In this section, we fix a prime p not equal to the characteristic of k_0. Our goal is to determine the invariants of PGL_p with values in $\mathbb{Z}/p\mathbb{Z}$.

The short exact sequence

$$1 \longrightarrow \boldsymbol{\mu}_n \longrightarrow \mathrm{SL}_n \longrightarrow \mathrm{PGL}_n \longrightarrow 1$$

gives a connecting homomorphism $\delta\colon H^1(k, \mathrm{PGL}_n) \to H^2(k, \boldsymbol{\mu}_n)$, cf. [**Se 02**, Ch. III], [**KMRT**, p. 386], or Example 1.2.2. We remark that δ has kernel zero because $H^1(k, \mathrm{SL}_n)$ is zero.

6.1. PROPOSITION. $\mathrm{Inv}^{\mathrm{norm}}_{k_0}(\mathrm{PGL}_p, \mathbb{Z}/p\mathbb{Z})$ *is a free $R_p(k_0)$-module with basis δ.*

The proposition will be proved at the end of this section. The reader is invited to compare this result with the examples of invariants of PGL_4 given in Example 1.2.5.

6.2. CYCLIC ALGEBRAS OF DEGREE n. Let n be a natural number not divisible by $\mathrm{char}\, k_0$ and fix a primitive n-th root of unity ζ in some separable closure of k_0. Let e_i denote the i-th standard basis vector of k_0^n. Define $u, v \in \mathrm{GL}_n$ to be the matrices such that

$$u(e_i) = \zeta^i e_i \text{ and } v(e_i) = \begin{cases} e_{i+1} & \text{for } 1 \leq i < n \\ e_1 & \text{for } i = n. \end{cases}$$

The maps $\mathbb{Z}/n\mathbb{Z} \to \mathrm{GL}_n$ and $\boldsymbol{\mu}_n \to \mathrm{GL}_n$ given by $i \mapsto v^i$ and $j \mapsto u^j$ are defined over k_0. Since $uv = \zeta vu$, there is a map

$$c\colon \mathbb{Z}/n\mathbb{Z} \times \boldsymbol{\mu}_n \to \mathrm{PGL}_n$$

defined over k_0 given by $(i, \zeta^j) \mapsto \overline{v}^i\overline{u}^j$ for $\overline{u}$, $\overline{v}$ the images of u, v in PGL_n.

The sets $H^1(k, \mathbb{Z}/n\mathbb{Z})$ and $H^1(k, \mathrm{PGL}_n)$ classify cyclic extensions k' of k and central simple k-algebras of degree n respectively. Recall that $H^1(k, \boldsymbol{\mu}_n) = k^\times/k^{\times n}$. *The map*

$$c_*\colon H^1(k, \mathbb{Z}/n\mathbb{Z}) \times H^1(k, \boldsymbol{\mu}_n) \to H^1(k, \mathrm{PGL}_n) \tag{6.3}$$

sends the cyclic extension k' and $\alpha \in k^\times/k^{\times n}$ to the class of the cyclic algebra (k', α), see Exercise 6.4 below.

6.4. EXERCISE. The cyclic algebra (k', α) is defined to be the k-algebra generated by k' and an element z such that $z\ell = \rho(\ell)z$ for all $\ell \in k'$ and ρ a fixed generator of $\mathrm{Gal}(k'/k)$. Justify the italicized claim in 6.2.

[One possible solution: Fix a separable closure k_{sep} of k. The image of k' and α under c_* define a 1-cocycle in $H^1(k, \mathrm{PGL}_n)$, which defines a twisted Galois action on $M_n(k_{\mathrm{sep}})$. A 1-cocycle determining k' also determines a preferred generator of $\mathrm{Gal}(k'/k)$; fix an element $\rho \in \mathrm{Gal}(k_{\mathrm{sep}}/k)$ which restricts to this preferred generator. Prove that the map $f : k' \to M_n(k_{\mathrm{sep}})$ given by $f(\beta)e_i = \rho^i(\beta)e_i$ is defined over k. Fix an n-th root a of α. Prove that the element $z = av^{-1}$ in $M_n(k_{\mathrm{sep}})$ is k-defined. Conclude that the fixed subalgebra of $M_n(k_{\mathrm{sep}})$ is isomorphic to (k', α).]

6.5. REMARK. The composition δc_* is a map

$$H^1(k, \mathbb{Z}/n\mathbb{Z}) \times H^1(k, \boldsymbol{\mu}_n) \to H^2(k, \boldsymbol{\mu}_n).$$

There is another such map given by the cup product; they are related by

$$\delta c_*(k', \alpha) = -(k')\cdot(\alpha),$$

see [**KMRT**, pp. 397, 415].

6.6. LEMMA. *If A is a central simple algebra over k of degree p, there is a finite extension k'/k of dimension prime to p over which A becomes cyclic.*

That is, the map (6.3) is surjective at p.

PROOF. This is well known. Recall the proof. We may assume that A is a division algebra, in which case it contains a field L that is a separable extension of k of dimension p. Let E be the smallest Galois extension of k containing L (in some algebraic closure of k); the Galois group Γ of E/k is a transitive subgroup of the symmetric group S_p; a p-Sylow subgroup S of Γ is thus cyclic of order p. Take for k' the subfield of E fixed by S. We have $E = Lk'$. Hence E is a cyclic extension of k' of dimension p which splits A over k'. $\square$

6.7. INVARIANTS OF A PRODUCT. Suppose we have algebraic groups G and G' such that

(6.8) There is a set $\{a_i\} \subset \mathrm{Inv}^{\mathrm{norm}}_{k_0}(G, C)$ that is an $R_n(k)$-basis of $\mathrm{Inv}^{\mathrm{norm}}_{k}(G, C)$ for every extension k/k_0, and

(6.9) There is an $R_n(k_0)$-basis $\{b_j\}$ of $\mathrm{Inv}_{k_0}(G', \boldsymbol{\mu}_n)$,

where n is the exponent of C. The cup product

$$H^{d_1}(*, C(d_1 - 1)) \times H^{d_2}(*, \mathbb{Z}/n\mathbb{Z}(d_2)) \to H^{d_1+d_2}(*, C(d_1 + d_2 - 1))$$

induces an $R_n(k_0)$-module homomorphism

$$\mathrm{Inv}^{\mathrm{norm}}_{k_0}(G, C) \otimes_{R_n(k_0)} \mathrm{Inv}_{k_0}(G', \boldsymbol{\mu}_n) \to \mathrm{Inv}^{\mathrm{norm}}_{k_0}(G \times G', C). \tag{6.10}$$

LEMMA. *The map (6.10) is injective. Its image I is the set of normalized invariants whose restriction to $H^1(*, G')$ is zero. The images of the $a_i \otimes b_j$ form a basis for I as an $R_n(k_0)$-module.*

This is a slight variation of Exercise 16.5 in S. We give a proof because we will use this result repeatedly later.

PROOF. Let c be a normalized invariant of $G\times G'$ with values in C that vanishes on $H^1(*,G')$. For a given k-G'-torsor T', the map

$$c_{T'}: T \mapsto c(T\times T')$$

is an invariant of G with values in C. As c vanishes on $H^1(*,G')$, $c_{T'}$ is normalized. By (6.8), $c_{T'}$ is the map $T\mapsto \sum_i \lambda_{i,T'}a_i(T)$ for uniquely determined $\lambda_{i,T'}\in R_n(k)$. The maps $T'\mapsto \lambda_{i,T'}$ are invariants of G' and belong to $\mathrm{Inv}_{k_0}(G',\boldsymbol{\mu}_n)$, which by (6.9) can be written uniquely as $\sum \lambda_{i,j}b_j$ for $\lambda_{i,j}\in R_n(k_0)$. This proves that c is the image of $\sum \lambda_{i,j}a_i\otimes b_j$, hence that the image of (6.10) includes every normalized invariant whose restriction to $H^1(*,G')$ is zero. As the reverse inclusion is trivial, we have proved the second sentence in the lemma.

The proof of the first sentence is similar. Suppose that the invariant

$$T\times T' \mapsto \sum_{i,j} \lambda_{i,j}\cdot b_j(T')\cdot a_i(T)$$

of $G\times G'$ is zero, where the $\lambda_{i,j}$ are in $R_n(k_0)$. For each k-G'-torsor T', we find that $\sum_j \lambda_{i,j}\cdot b_j(T')$ is zero by (6.8), hence the invariant $\sum_j \lambda_{i,j}\cdot b_j$ is zero. By (6.9), $\lambda_{i,j}$ is zero for all i,j.

Because $\mathrm{Inv}^{\mathrm{norm}}_{k_0}(G,C)$ and $\mathrm{Inv}_{k_0}(G',\boldsymbol{\mu}_n)$ are free $R_n(k_0)$-modules, the third sentence in the lemma follows from the first two. □

In the examples below, the set $\{b_j\}$ is a basis of $\mathrm{Inv}_k(G',\boldsymbol{\mu}_n)$ for every extension k/k_0. This implies that the lemma holds when k_0 is replaced with k.

We can now prove Prop. 6.1.

PROOF OF PROP. 6.1. Combining Lemmas 6.6 and 5.3, we find that the map

$$c^*: \mathrm{Inv}^{\mathrm{norm}}_{k_0}(\mathrm{PGL}_p,\mathbb{Z}/p\mathbb{Z}) \to \mathrm{Inv}^{\mathrm{norm}}_{k_0}(\mathbb{Z}/p\mathbb{Z}\times\boldsymbol{\mu}_p,\mathbb{Z}/p\mathbb{Z}) \tag{6.11}$$

induced by c is an injection.

It follows from Remark 6.5 or [**KMRT**, 30.6] that $c_*(x,1)$ and $c_*(1,y)$ are the neutral class in $H^1(k,\mathrm{PGL}_p)$ for every extension k/k_0, every $x\in H^1(k,\mathbb{Z}/p\mathbb{Z})$, and every $y\in H^1(k,\boldsymbol{\mu}_p)$. In particular the image of (6.11) is contained in the submodule I of invariants that are zero on $H^1(*,\boldsymbol{\mu}_p)$.

Fix a normalized invariant a in $\mathrm{Inv}_{k_0}(\mathrm{PGL}_p,\mathbb{Z}/p\mathbb{Z})$. By Lemma 6.7, Prop. 3.7, and Prop. 2.1, its image c^*a under (6.11) is of the form

$$(x,y)\mapsto \lambda_1\cdot x+\lambda_2\cdot y\cdot x \qquad (x\in H^1(k,\mathbb{Z}/p\mathbb{Z}), y\in H^1(k,\boldsymbol{\mu}_p))$$

for uniquely determined $\lambda_1,\lambda_2\in R_p(k_0)$. But

$$(c^*a)(x,1)=a(c_*(x,1))=a(M_p(k))=0$$

for every $x\in H^1(k,\mathbb{Z}/p\mathbb{Z})$ and every extension k, so λ_1 is zero. Therefore,

$$(c^*a)(x,y)=\lambda_2\cdot y\cdot x.$$

Since c^*a is an $R_p(k_0)$-multiple of $c^*\delta$, we conclude that δ spans the module $\mathrm{Inv}^{\mathrm{norm}}(\mathrm{PGL}_p,\mathbb{Z}/p\mathbb{Z})$. □

6.12. REMARK. A versal torsor for $\mathbb{Z}/p\mathbb{Z}\times\boldsymbol{\mu}_p$ gives a PGL_p-torsor T. The injectivity of (6.11) combined with S12.3 shows that invariants a,a' of PGL_p that agree on T are the same. One may view T as a "p-versal torsor" (appropriately defined) for PGL_p.

6.13. Open problem. (Reichstein-Youssin [**RY**, p. 1047]) Let k_0 be an algebraically closed field of characteristic zero. Does there exist a nonzero invariant $H^1(*, PGL_{p^r}) \to H^{2r}(*, \mathbb{Z}/p\mathbb{Z})$?

[For $p = r = 2$, one has the Rost-Serre-Tignol invariant described in Example 1.2.5.]

6.14. Question. Let k_0 be an algebraically closed field of characteristic zero. What are the mod 2 invariants of PGL_4? That is, what is $\mathrm{Inv}_{k_0}^{\mathrm{norm}}(\mathrm{PGL}_4, \mathbb{Z}/2\mathbb{Z})$?

[This is a "question" and not an "exercise" because there are central simple algebras of degree 4 that are neither cyclic nor tensor products of two quaternion algebras [**Alb**].]

7. Extending invariants

7.1. Fix functors A and A' mapping $\mathsf{Fields}_{/k_0} \to \mathsf{Sets}$ and a morphism $\phi : A' \to A$. *When can an invariant $a' : A' \to M(*, C)$ be extended to an invariant $a : A \to M(*, C)$?* That is, when is there an invariant a that makes the diagram

$$\begin{array}{ccc} A' & \xrightarrow{a'} & M(*, C) \\ {\scriptstyle\phi}\downarrow & \nearrow_{a} & \\ A & & \end{array}$$

commute?

Clearly, we must have

$$\text{(7.2)} \qquad \begin{array}{l} \text{For every extension } k/k_0 \text{ and every } x, y \in A'(k): \\ \qquad \phi(x) = \phi(y) \quad \Longrightarrow \quad a'(x) = a'(y) \end{array}$$

Proposition. *If ϕ satisfies* (5.2), *then the restriction*

$$\phi^* : \mathrm{Inv}_{k_0}(A, C) \to \mathrm{Inv}_{k_0}(A', C)$$

defines an isomorphism of $\mathrm{Inv}_{k_0}(A, C)$ *with the set of invariants a' of A' satisfying* (7.2).

That is, assuming (5.2), condition (7.2) is sufficient as well as necessary.

Note that the proposition gives a solution to Exercise 22.9 in S as a corollary. That is, if ϕ satisfies (5.2) and ϕ is injective, then *the restriction map*

$$\phi^* : \mathrm{Inv}_{k_0}(A, C) \to \mathrm{Inv}_{k_0}(A', C)$$

is an isomorphism.

The rest of this section is a proof of the proposition. The homomorphism ϕ^* is injective by Lemma 5.3, so it suffices to prove that every invariant a' of A' satisfying (7.2) is in the image. As in 3.1, we may assume that the exponent of C is the power of a prime p.

7.3. For each *perfect* field k/k_0 and each $x \in A(k)$, we define an element $a(x) \in M(k, C)$ as follows. Fix an extension k_2 of k as in 5.1, i.e., such that there is an $x' \in A'(k_2)$ such that $\phi(x')$ is the restriction of x and the dimension $[k_2 : k]$ is not divisible by p.

Lemma A. *$a'(x')$ is the restriction of a unique element of $M(k, C)$.*

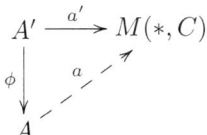

We define $a(x)$ to be the unique element of $M(k, C)$ such that $\operatorname{res}_{k_2/k} a(x)$ is $a'(x')$. For the proof of this lemma and Lemma B below, we fix a separable closure k_{sep} of k_2 (hence also of k).

PROOF OF LEMMA A. Uniqueness is easy, so we prove that $a'(x')$ is defined over k.

For each finite extension k_3 of k_2 in k_{sep} and every $\sigma \in \operatorname{Gal}(k_{\text{sep}}/k)$ such that $\sigma(k_3) \supseteq k_2$, we claim that

$$\sigma_* \operatorname{res}_{k_3/k_2} a'(x') = \operatorname{res}_{\sigma(k_3)/k_2}(a'(x')) \tag{7.4}$$

in $M(\sigma(k_3), C)$, i.e., that $a'(x')$ is "stable" in $M(k_2, C)$. Indeed, the equation

$$\sigma_* \operatorname{res}_{k_3/k_2} x = \operatorname{res}_{\sigma(k_3)/k_2} x,$$

holds because x is defined over k. This implies

$$\phi(\sigma_* \operatorname{res}_{k_3/k_2} x') = \phi(\operatorname{res}_{\sigma(k_3)/k_2} x')$$

because ϕ is defined over k and so commutes with σ_* and restriction. By (7.2), we have:

$$a'(\sigma_* \operatorname{res}_{k_3/k_2} x') = a'(\operatorname{res}_{\sigma(k_3)/k_2} x')$$

But the invariant a' is also defined over k, so it commutes with σ_* and restriction. We conclude that equation (7.4) holds.

Combining (7.4) with the double coset formula for the composition $\operatorname{res} \circ \operatorname{cor}$ as in [**AM**, Th. II.6.6] shows that $a'(x')$ is the restriction of an element of $M(k, C)$. □

LEMMA B. *The element $a(x) \in M(k, C)$ depends only on x* (and not on the choice of k_2 and x').

PROOF. Let ℓ_2 be a finite extension of k in k_{sep} such that $\operatorname{res}_{\ell_2/k} x$ is the image of some $y' \in A'(\ell_2)$ and the prime p does not divide $[\ell_2 : k]$. (I.e., ℓ_2 is an extension as provided by 5.1, and it is separable because k is perfect.) We prove that $a'(x') \in M(k_2, C)$ and $a'(y') \in M(\ell_2, C)$ are restrictions of the same element in $M(k, C)$.

Case 1: ℓ_2 is a conjugate of k_2. Suppose that there is a $\sigma \in \operatorname{Gal}(k_{\text{sep}}/k)$ such that $\sigma(\ell_2)$ equals k_2. One quickly checks that $\phi(\sigma_* y')$ equals $\phi(x')$ in $A(k_2)$, hence $a'(x')$ equals $a'(\sigma_* y')$ by (7.2), i.e., $a'(x')$ is $\sigma_* a'(y')$. The lemma follows in this special case.

Case 2. Suppose that the compositum K of k_2 and ℓ_2 in k_{sep} has dimension $[K : k]$ not divisible by the prime p. Since $\phi(\operatorname{res}_{K/k_2} x')$ and $\phi(\operatorname{res}_{K/\ell_2} y')$ equal $\operatorname{res}_{K/k}(x)$, the restriction of $a'(x')$ and $a'(y')$ in K agree. By the hypothesis on the dimension $[K : k]$, the lemma holds in this special case.

Case 3: general case. Let S be a p-Sylow in $\operatorname{Gal}(k_{\text{sep}}/k)$ fixing k_2 elementwise. There is a $\sigma \in \operatorname{Gal}(k_{\text{sep}}/k)$ such that $\sigma(\ell_2)$ is also fixed elementwise by S. It follows that the compositum of $\sigma(\ell_2)$ and k_2 has dimension over K not divisible by p. A combination of cases 1 and 2 gives the lemma in the general case. □

7.5. For an arbitrary extension k of k_0, write k_p for the "perfect closure" of k. Since $M(k, C)$ is canonically isomorphic to $M(k_p, C)$, we define $a(x)$ to be the element $a(\operatorname{res}_{k_p/k} x) \in M(k_p, C)$ defined in 7.3 above.

For every extension k of k_0, we have defined a function a_k making the diagram

$$\begin{array}{ccc} A'(k) & \xrightarrow{a'_k} & M(k,C) \\ {\scriptstyle \phi_k}\big\downarrow & \nearrow_{a_k} & \\ A(k) & & \end{array}$$

commute. We leave the proof that this defines a morphism of functors $a : A \to M(*, C)$ to the reader.

8. Mod 3 invariants of Albert algebras

In this section, we assume that k_0 has characteristic $\neq 2, 3$ and classify the normalized mod 3 invariants of Albert algebras. Recall that Albert k-algebras are 27-dimensional exceptional Jordan algebras—see [**SV**, Ch. 5], [**PeR 94**], or [**KMRT**, Ch. IX]—and we write Alb for the functor such that $\mathrm{Alb}(K)$ is the isomorphism classes of Albert K-algebras. We compute $\mathrm{Inv}^{\mathrm{norm}}(\mathrm{Alb}, \mathbb{Z}/3\mathbb{Z})$.

8.1. Example. Let $M = M_3(k)$ be the algebra of 3-by-3 matrices over k. On the 27-dimensional space $J = M \times M \times M$, define a cubic form N by

$$N(a, b, c) = \det(a) + \det(b) + \det(c) - \mathrm{tr}(abc).$$

Write 1 for the element $(1, 0, 0)$ in J. The "Springer construction" endows J with the structure of an Albert k-algebra induced by N and the choice of the element 1, see [**McC**, §5]. It is the *split* Albert algebra and its automorphism group F_4 is the subgroup of $\mathrm{GL}(J)$ consisting of elements that fix 1 and N [**J 59**, Th. 4]; it is a split algebraic group of type F_4. By Galois descent we have an isomorphism of functors $H^1(*, F_4) \cong \mathrm{Alb}(*)$, see [**KMRT**, p. 517]. This isomorphism identifies $\mathrm{Inv}^{\mathrm{norm}}(\mathrm{Alb}, \mathbb{Z}/3\mathbb{Z})$ with $\mathrm{Inv}^{\mathrm{norm}}(F_4, \mathbb{Z}/3\mathbb{Z})$.

8.2. If (g, z) is a point of $\mathrm{PGL}_3 \times \boldsymbol{\mu}_3$, let $t(g, z)$ be the element of $\mathrm{GL}(J)$ defined by

$$(a, b, c) \mapsto (i_g(a), z \cdot i_g(b), z^2 \cdot i_g(c)),$$

where i_g is the inner autormorphism of M defined by g. Since (g, z) fixes both 1 and N, it belongs to the group F_4. This gives an inclusion $t \colon \mathrm{PGL}_3 \times \boldsymbol{\mu}_3 \to F_4$ and a corresponding map

$$t_* \colon H^1(*, \mathrm{PGL}_3) \times H^1(*, \boldsymbol{\mu}_3) \to H^1(*, F_4) \cong \mathrm{Alb}(*). \tag{8.3}$$

The image of a pair (A, α) is often denoted by $J(A, \alpha)$; such algebras are known as *first Tits constructions*, cf. [**KMRT**, §39.A].

Every Albert k-algebra is a first Tits construction or becomes one over a quadratic extension of k—see, e.g., [**KMRT**, 39.19]—so the map in (8.3) is surjectve at every odd prime and the restriction map

$$t^* \colon \mathrm{Inv}^{\mathrm{norm}}(F_4, \mathbb{Z}/3\mathbb{Z}) \to \mathrm{Inv}^{\mathrm{norm}}(\mathrm{PGL}_3 \times \boldsymbol{\mu}_3, \mathbb{Z}/3\mathbb{Z}) \tag{8.4}$$

is injective by Lemma 5.3.

8.5. Invariants of F_4 mod 3. Consider the invariant

$$g_3 \colon H^1(*, \mathrm{PGL}_3) \times H^1(*, \boldsymbol{\mu}_3) \to H^3(*, \boldsymbol{\mu}_3^{\otimes 2})$$

defined by $g_3(A, \alpha) = \delta(A) \cdot (\alpha)$ for δ as defined in §6. We now give two arguments that g_3 is the restriction of an invariant of F_4.

PROOF #1. The meat of [**PeR 96**] is their Lemma 4.1, which says that g_3 "factors through" the image of (8.3) in $H^1(*, F_4)$. That is, if the first Tits constructions $J(A, \alpha)$ and $J(A', \alpha')$ are isomorphic, then $g_3(A, \alpha)$ equals $g_3(A', \alpha')$. Prop. 7.1 gives that g_3 extends to an invariant of F_4. □

PROOF #2. The Dynkin index of F_4 is 6 [**Mer**, 16.9], so the mod 3 portion of the Rost invariant gives a nonzero invariant

$$g_3' \colon H^1(*, F_4) \to H^3(*, \boldsymbol{\mu}_3^{\otimes 2}).$$

Applying Lemma 6.7, we conclude that $t^* g_3'$ equals λg_3 for some fixed $\lambda \in R_3(k_0)$. Since the image of g_3' under $\mathrm{Inv}_{k_0}(G, \mathbb{Z}/3\mathbb{Z}) \to \mathrm{Inv}_K(G, \mathbb{Z}/3\mathbb{Z})$ is nonzero for every extension K/k_0, the invariant $t^* g_3'$ is nonzero over every K, and we conclude that $\lambda = \pm 1$, i.e., $t^* g_3'$ is $\pm g_3$. □

We abuse notation by writing g_3 also for the invariant $(t^*)^{-1}(g_3)$ of F_4. This invariant was originally constructed in [**Rost 91**].

8.6. PROPOSITION. $\mathrm{Inv}_{k_0}^{\mathrm{norm}}(F_4, \mathbb{Z}/3\mathbb{Z})$ *is a free* $R_3(k_0)$*-module with basis* g_3.

PROOF. We imitate the proof of Prop. 6.1, with the role of $\mathbb{Z}/p\mathbb{Z} \times \boldsymbol{\mu}_p$ played by $\mathrm{PGL}_3 \times \boldsymbol{\mu}_3$. For every central simple algebra A over every extension k/k_0 and every $\alpha \in k^\times$, the algebra $J(A, \alpha)$ is "split", i.e., $t_*(A, \alpha)$ is the neutral class in $H^1(k, F_4)$, if and only if α is the reduced norm of an element of $A^\times$ by [**J 68**, p. 416, Th. 20] or [**McC**, Th. 6]. In particular, $t_*(M_3(k), \alpha)$ is the neutral class for every α, and Lemma 6.7 gives that the restriction of a normalized invariant in $\mathrm{Inv}_{k_0}(F_4, \mathbb{Z}/3\mathbb{Z})$ to $\mathrm{PGL}_3 \times \boldsymbol{\mu}_3$ can be written as

$$(A, \alpha) \mapsto \lambda_1 \cdot [A] + \lambda_2 \cdot (\alpha) \cdot [A]$$

for uniquely determined $\lambda_1, \lambda_2 \in R_3(k_0)$. But the algebra $J(A, 1)$ is also split for every A. It follows that λ_1 is zero. This proves that g_3 spans $\mathrm{Inv}_{k_0}^{\mathrm{norm}}(F_4, \mathbb{Z}/3\mathbb{Z})$. □

Combining the proposition with the classification of the invariants mod 2 in S22.5, we have found just three interesting invariants of F_4, namely g_3, f_3, and f_5. Perhaps the outstanding open problem in the theory of Albert algebras is:

8.7. OPEN PROBLEM. (Serre [**Se 95**, §9.4], [**PeR 94**, Q. 1, p. 205]) Is the map

$$g_3 \times f_3 \times f_5 \colon H^1(*, F_4) \to H^3(*, \mathbb{Z}/3\mathbb{Z}) \times H^3(*, \mathbb{Z}/2\mathbb{Z}) \times H^5(*, \mathbb{Z}/2\mathbb{Z})$$

injective? That is, is an Albert algebra J determined up to isomorphism by its invariants $g_3(J)$, $f_3(J)$, and $f_5(J)$?

[The map is injective on the kernel of g_3 [**SV**, 5.8.1], i.e., "reduced Albert algebras are classified by their trace form". Also, Rost has an unpublished result on this problem, see [**Rost 02**]. Note that it is still unknown if the map is injective on the kernel of $f_3 \times f_5$, i.e., for first Tits constructions.]

One can also ask about the image of the map $g_3 \times f_3 \times f_5$. Let $k = \mathbb{Q}_p(t)$ for some prime p. Serre points out in [**Se 95**, 9.5] that for every Albert k-algebra A, it is not possible that $f_3(A)$ and $g_3(A)$ are both non-zero. That is, the cohomology mod 2 and the cohomology mod 3 are somehow related.

8.8. SYMBOLS. We now drop the assumption that the characteristic of k_0 is $\neq 2, 3$, and instead assume that it does not divide some fixed natural number n.

We call an element $x \in H^d(k, \boldsymbol{\mu}_n^{\otimes(d-1)}) = M^d(k, \mathbb{Z}/n\mathbb{Z})$ (for $d \geq 2$) a *symbol* if it is in the image of the cup product map

$$H^1(k, \mathbb{Z}/n\mathbb{Z}) \times \underbrace{H^1(k, \boldsymbol{\mu}_n) \times \cdots \times H^1(k, \boldsymbol{\mu}_n)}_{d-1 \text{ copies}} \to H^d(k, \boldsymbol{\mu}_n^{\otimes(d-1)}).$$

In particular, the zero class is always a symbol. In the usual identification of $H^2(k, \boldsymbol{\mu}_n)$ with the n-torsion in the Brauer group of k, symbols are identified with cyclic algebras of degree n as defined in §6.

8.9. EXAMPLE. In the case $n = 2$, $M^d(k, \mathbb{Z}/2\mathbb{Z})$ is just $H^d(k, \mathbb{Z}/2\mathbb{Z})$, and it is isomorphic to I^d/I^{d-1} as in 1.2.3. Symbols in $M^d(k, \mathbb{Z}/2\mathbb{Z})$ correspond to the (equivalence classes of) d-Pfister quadratic forms. Further, one has the following nice property: If there is an odd-dimensional extension K/k such that $\mathrm{res}_{K/k}(x)$ is a symbol in $H^d(k, \mathbb{Z}/2\mathbb{Z})$, then x is itself a symbol by [**Rost 99a**, Prop. 2].

In the case $n = 3$ (and $\operatorname{char} k_0 \neq 3$), we have the following weaker property, mentioned in [**Rost 99a**]:

8.10. LEMMA. *Fix $x \in H^2(k, \boldsymbol{\mu}_3)$. If there is an extension K/k such that 3 does not divide the dimension $[K : k]$ and $\mathrm{res}_{K/k}(x)$ is a symbol in $H^2(K, \boldsymbol{\mu}_3)$, then x is itself a symbol.*

PROOF. We identify $H^2(k, \boldsymbol{\mu}_3)$ and $H^2(K, \boldsymbol{\mu}_3)$ with the 3-torsion in the Brauer group of k and K respectively. We assume that x is nonzero, hence that it corresponds to a central division k-algebra A of degree 3^r for some positive r. By hypothesis, $A \otimes K$ is isomorphic to $M_r(B)$ for a cyclic K-algebra B of degree 3. But as 3 does not divide $[K : k]$, the index of A and $A \otimes K$ agree [**Dr**, §9, Th. 12]. It follows that A is a division algebra of degree 3 over k, hence by Wedderburn's Theorem [**KMRT**, 19.2] A is cyclic, i.e., x is a symbol. □

Returning to groups of type F_4, the image of the invariant

$$g_3 \colon H^1(k, F_4) \to H^3(k, \boldsymbol{\mu}_3^{\otimes 2})$$

consists of symbols by [**Th**, p. 303]. For an alternative proof, combine [**KMRT**, 40.9] with Lemma 8.10.

Part II

Surjectivities and invariants of E_6, E_7, and E_8

9. Surjectivities: internal Chevalley modules

Consider the following:

9.1. EXAMPLE. Let q be a nondegenerate quadratic form on a vector space V over a field k of characteristic $\neq 2$. Fix an anisotropic vector $v \in V$. Over a separable closure k_{sep} of k, the orthogonal group $\mathrm{O}(q)(k_{\mathrm{sep}})$ acts transitively on the open subset of $\mathbb{P}(V)$ consisting of anisotropic vectors by Witt's Extension Theorem. The stabilizer of an anisotropic line $[v]$ in $\mathrm{O}(q)$ is isomorphic to $\boldsymbol{\mu}_2 \times \mathrm{O}(v^\perp)$. It follows from [**Se 02**, §I.5.5, Prop. 37] that the natural map

$$H^1(k, \boldsymbol{\mu}_2 \times \mathrm{O}(v^\perp)) \to H^1(k, \mathrm{O}(q))$$

is surjective. Repeating this procedure, we find a *surjection*

$$\bigoplus^{\dim V} H^1(k, \boldsymbol{\mu}_2) \to H^1(k, \mathrm{O}(q)). \tag{9.2}$$

The set $H^1(k, \mathrm{O}(q))$ classifies nondegenerate quadratic forms on V and $H^1(k, \boldsymbol{\mu}_2)$ is the same as $k^\times/k^{\times 2}$. Tracing through these identifications, we find that the map (9.2) sends a tuple $\alpha_1, \alpha_2, \ldots, \alpha_n$ of elements of $k^\times/k^{\times 2}$ to the diagonal quadratic form $\langle \alpha_1, \alpha_2, \ldots, \alpha_n \rangle$. That is, we have found a high-tech proof of the fact that quadratic forms can be diagonalized.

(Some readers might object that this is not really a different proof, because we used Witt's Extension Theorem. But we only used it to prove the transitivity of the action over a separable closure of k, and this could have been deduced from the general machinery presented in 9.9 and 9.14 below.)

9.3. Let G be an algebraic group over k. Roughly speaking, we now abstract the preceding example by finding a subgroup N of G such that the natural map $H^1(*, N) \to H^1(*, G)$ is surjective. We suppose that k is infinite[h] and that G has a representation V such that there is an open G-orbit in $\mathbb{P}(V)$ over an algebraic closure of k. As k is infinite, there is a k-point $[v]$ in the open orbit.

THEOREM. *The natural map*

$$H^1_{\mathrm{fppf}}(k, N) \to H^1(k, G)$$

is surjective, where N is the scheme-theoretic stabilizer of $[v]$ in G.

We write $H^1_{\mathrm{fppf}}(k, N)$ for the pointed set of k-N-torsors relative to the fppf topology as in [**DG 70**]. When N is smooth, this group agrees with the usual Galois cohomology set $H^1(k, N)$ [**DG 70**, p. 406, III.5.3.6], so the reader who wishes to avoid flat cohomology may simply add hypotheses that various groups are smooth or—more restrictively—only consider fields of characteristic zero.

In the case where N is smooth, a concrete proof of the theorem can be found in [**Ga 01a**, 3.1] or by applying [**Se 02**, §III.2.1, Exercise 2] with B, C, D replaced by $G, N, GL(V)$.

PROOF. Write $\mathcal{O}$ for the G-orbit of $[v]$ in $\mathbb{P}(V)$ (equivalently, G/N). For $z \in H^1(k, G)$, there is an inclusion of twisted objects $\mathcal{O}_z \to \mathbb{P}(V)_z$. As G acts on $\mathbb{P}(V)$ through $\mathrm{GL}(V)$, the twisted variety $\mathbb{P}(V)_z$ is isomorphic to $\mathbb{P}(V)$ and the k-points are dense in $\mathbb{P}(V)_z$ (because k is infinite). Moreover, $\mathcal{O}_z$ is open in $\mathbb{P}(V)_z$ because

[h] This hypothesis is harmless. In the examples, G will be connected, so $H^1(k, G)$ will be zero when k is finite.

$\mathcal{O}$ is open in $\mathbb{P}(V)$. Hence $\mathcal{O}_z$ has a k-point and z is in the image of the map $H^1_{\text{fppf}}(k, N) \to H^1(k, G)$ by [**DG 70**, p. 373, Prop. III.4.4.6b] (an fppf analogue of Prop. 37 in [**Se 02**]). □

9.4. EXAMPLE (char $k = 0$). Let G be a semisimple group. *The adjoint representation V of G has an open orbit in $\mathbb{P}(V)$ if and only if G has absolute rank 1*, i.e., G is of type A_1. Essentially, this is because the regular semisimple elements in V are an open subvariety. Indeed, if G is of rank 1, then the regular semisimple elements in V are actually an open orbit in $\mathbb{P}(V)$, because G acts transitively on the collection of maximal toral subalgebras of V [**Hu**, 16.4]. Conversely, if there is an open G-orbit in $\mathbb{P}(V)$, it contains $[v]$ for some regular semisimple element v. This v is contained in a maximal toral subalgebra of V, and by conjugacy of tori, we may assume that this subalgebra is the Lie algebra of a maximal torus T in G. Then T fixes v, so it is contained in the stabilizer N of $[v]$. We have:

$$1 \le \operatorname{rank} G \le \dim N = \dim G - \dim \mathbb{P}(V) = 1.$$

9.5. A non-example is furnished by a representation V of G on which G acts trivially. If $\dim V = 1$, then $\mathbb{P}(V)$ is a point, N equals G, and the conclusion of Th. 9.3 is uninteresting. If $\dim V$ is at least 2, then $\mathbb{P}(V)$ does not have an open orbit (exercise).

9.6. EXAMPLE (Reducible representations). Let V be a representation of G as in 9.3, and suppose that there is a proper G-invariant subspace W of V. The quotient map $V \to V/W$ gives a G-equivariant rational surjection $f\colon \mathbb{P}(V) \dashrightarrow \mathbb{P}(V/W)$. If $[v]$ is in the open G-orbit in $\mathbb{P}(V)$, then v is not in W, the map f is defined at $[v]$ and the orbit of $f([v])$ is dense in $\mathbb{P}(V/W)$, hence open.

9.7. EXAMPLE (char $k = 0$). A group G of type E_8 has no nontrivial representations V with an open G-orbit in $\mathbb{P}(V)$. Indeed, by Example 9.6, it suffices to prove that no faithful *irreducible* representation V of G has an open orbit in $\mathbb{P}(V)$. By the following exercise, the constraint $\dim G \ge \mathbb{P}(V)$ leaves the adjoint representation as the only possibility, and there is no open G-orbit in that case by Example 9.4.

9.8. EXERCISE (char $k = 0$). Check that the split group of type E_8 has unique irreducible representations of dimensions 1, 248 (adjoint), 3875, 27000, and 30380, and no others of dimension $< 10^5$.

[Compare [**McKP**, p. 79] or Sequence A121732 in [**Sl**].]

9.9. INTERNAL CHEVALLEY MODULES. How to find groups and representations that satisfy the hypotheses of Theorem 9.3? We now give a mechanism from representation theory that produces such.

Let $\widetilde{G}$ be a semisimple algebraic group that is defined and isotropic over k. We fix a maximal k-torus $\widetilde{T}$ in $\widetilde{G}$ that contains a maximal k-split torus $\widetilde{T}_d$. Fix also a set $\widetilde{\Delta}$ of simple roots of $\widetilde{G}$ with respect to $\widetilde{T}$. We suppose that there is some $\pi \in \widetilde{\Delta}$ that is fixed by the Galois group (under the $*$-action, which permutes $\widetilde{\Delta}$) and is not constant on $\widetilde{T}_d$. (In the notation of Tits's classification paper [**Tits 66**], the vertex π in the Dynkin diagram is circled and the circle does not include any other vertices.) Finally, we assume that k has characteristic $\neq 2$ if $\widetilde{G}$ is of type B, C, or F_4 and $\neq 2, 3$ if $\widetilde{G}$ has type G_2. This concludes our list of assumptions.

We define G to be the semisimple subgroup of $\widetilde{G}$ that is generated over a separable closure k_{sep} of k by the 1-dimensional unipotent subgroups U_α of $\widetilde{G}$ as α varies over the roots of $\widetilde{G}$ with π-coordinate zero. The Dynkin diagram of G is the diagram of $\widetilde{G}$ with the vertex π deleted. If $\widetilde{G}$ is simply connected, then so is G by [**SS**, 5.4b]. The reader can find a list of Dynkin diagrams in Table 9 below and a list of concrete examples of groups G that we will consider by looking ahead at Tables 23a or 12.

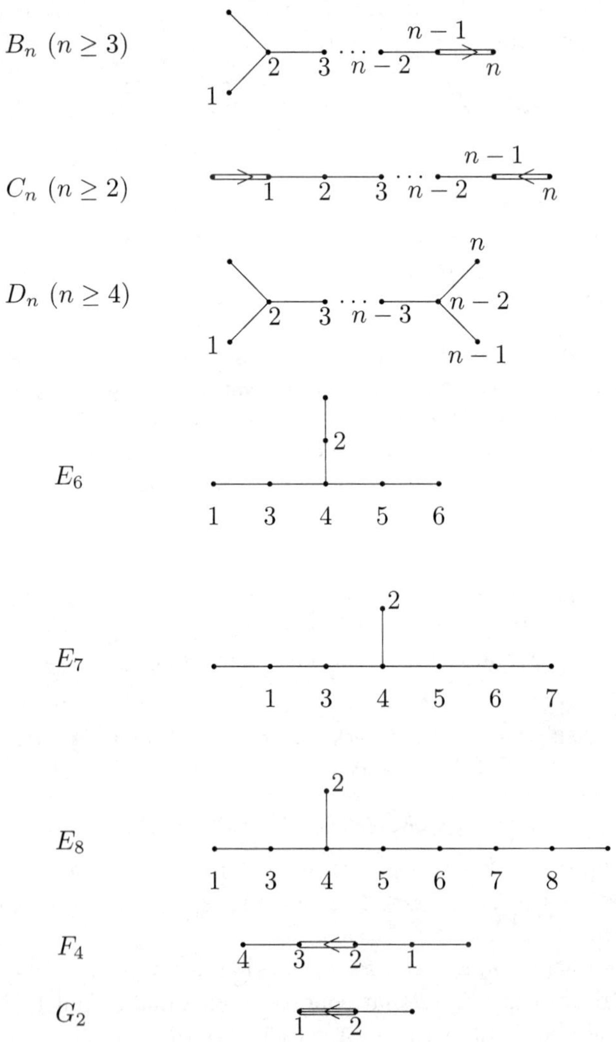

TABLE 9. Extended Dynkin diagrams

Vertices are numbered as in [**Bou Lie**]. The unlabeled vertex corresponds to the negative $-\widetilde{\alpha}$ of the highest root, and omitting this vertex leaves the usual Dynkin diagram. Type A is omitted entirely.

Over k_{sep}, there is a cocharacter $\lambda\colon \mathbb{G}_m \to \widetilde{T}$ such that $\pi(\lambda)$ is negative and $\alpha(\lambda)$ is zero for $\alpha \in \widetilde{\Delta} \setminus \{\pi\}$. The cocharacter λ is even defined over k by [**BT**, 6.7, 6.9]; its image is in $\widetilde{T}_d$. We take $\widetilde{P}$ (respectively, L) to be the parabolic subgroup of $\widetilde{G}$ (resp., Levi subgroup of $\widetilde{P}$) picked out by λ in the sense of [**Sp 98**, 13.4.1], i.e., the subgroup generated over k_{sep} by $\widetilde{T}$ and the U_α where α varies over the roots of $\widetilde{G}$ with non-positive π-coordinate (resp., π-coordinate zero). Note that G is the derived subgroup of L.

The Levi subgroup L acts on the unipotent radical Q of $\widetilde{P}$ by conjugation. We fix a positive integer i and write $Q(i)$ for the subgroup of Q spanned by the U_α where the π-coordinate of α is $\leq -i$. We put $V := Q(1)/Q(2)$; it is a representation of L and there is an open L-orbit in V over an algebraic closure of k [**ABS**, Th. 2]. The representation V is called an *internal Chevalley module*. It is irreducible with highest weight $-\pi$ [**ABS**, Th. 2].

9.10. REMARKS. (1) The addition on V comes from the multiplication in $\widetilde{G}$. What is the scalar multiplication that turns V into a k-vector space? Suppose that $\widetilde{T}$ is split. Number the roots of $\widetilde{G}$ with π-coordinate -1 arbitrarily as $\rho_1, \rho_2, \ldots, \rho_s$. The product map

$$U_{\rho_1} \times U_{\rho_2} \times \cdots \times U_{\rho_s} \xrightarrow{m} V$$

is an isomorphism by [**Bor**, Prop. 14.4(2)]. The group U_α is the image of a homomorphism $x_\alpha\colon \mathbb{G}_a \to \widetilde{G}$ and the scalar multiplication is the naive one: For $\lambda \in k^\times$ and $u_i \in k$, we have

$$\lambda \cdot m\left(\prod x_{\rho_i}(u_i)\right) = m\left(\prod x_{\rho_i}(\lambda u_i)\right),$$

see [**ABS**, p. 554].

(2) If, instead of the parabolic $\widetilde{P}$, we chose the "opposite" parabolic, then everything would work out the same except that the highest weight of V would be the highest positive root with π-coordinate 1—something that is more difficult to read off of the Dynkin diagrams. The resulting V would be the L-module that is dual to the one we consider here.

(3) In most of the examples considered below, the vertex π of the Dynkin diagram is adjacent to only one other vertex—call it δ—and the two vertices are joined by a single bond, so the highest weight of V is the fundamental weight corresponding to δ.

(4) Although one could consider the modules $Q(i)/Q(i+1)$ for various i, no real generality is gained, see [**Rö 93c**, 1.8].

(5) Although the root π is fixed by the Galois group under the $*$-action (and the cocharacter λ is k-defined), π need not be fixed by the usual Galois action. Indeed, [**Ga 98**] gives a concrete construction of groups of type ${}^3D_4$ with k-rank 1 where the root $\pi := \alpha_2$ is fixed by the $*$-action (and is non-constant on the split torus), but the usual Galois action interchanges α_2 and $\alpha_1 + \alpha_2 + \alpha_3 + \alpha_4$.

(6) In case $\widetilde{G}$ is split, there is an open $\widetilde{P}$-orbit in the unipotent radical Q whose elements are known as "Richardson elements". Clearly, any Richardson element with π-coordinate -1 maps to an element of the open L-orbit in V. For $\widetilde{G}$ of classical type, the reader can find concrete examples of Richardson elements in [**Ba**].

9.11. We maintain the assumptions from 9.9, and we further assume—as in 9.3—that the field k is infinite. We fix a k-point $[v]$ in the open L-orbit in $\mathbb{P}(V)$.

THEOREM. *The natural map*

$$H^1_{\text{fppf}}(k, N) \to H^1(k, G)$$

is surjective, where N is the scheme-theoretic stabilizer of $[v] \in \mathbb{P}(V)$ in G.

PROOF. Write $\mathcal{O}$ for the L-orbit of $[v]$ in $\mathbb{P}(V)$. Note that since the L-orbit of v is dense in V, $\mathcal{O}$ is dense in $\mathbb{P}(V)$, hence open in $\mathbb{P}(V)$ because orbits are locally closed.

As V is an irreducible representation of L, the torus S in the center of L acts on V by scalar multiplication. But G and S generate L, so the G- and L-orbits in $\mathbb{P}(V)$ coincide. That is, the G-orbit of $[v] \in \mathbb{P}(V)$ is open. Theorem 9.3 completes the proof. $\square$

Theorems 9.3 and 9.11 give surjections $H^1_{\text{fppf}}(K, N) \to H^1(K, G)$ for every extension K/k, where N is the scheme-theoretic stabilizer of a k-point in the open orbit. We summarize this by saying that the morphism of functors $H^1_{\text{fppf}}(*, N) \to H^1(*, G)$ *is surjective* or that the inclusion $N \subset G$ *induces a surjection on H^1's.*

9.12. EXAMPLE ($F_4 \times \boldsymbol{\mu}_3 \subset E_6$). The natural inclusion of root systems leads to an inclusion of split simply connected groups $E_6 \subset E_7$. We take these groups as G and $\widetilde{G}$ respectively in the notation of 9.11, so that π is the root α_7 of E_7. (We number the simple roots as in Table 9.) The representation V of G is irreducible and 27-dimensional with highest weight the fundamental dominant weight corresponding to α_6.

There is a split group of type F_4 inside of E_6, and we denote it also by F_4. Writing $x_\alpha : \mathbb{G}_a \to E_6$ for the generators of E_6 as in [**St**], F_4 is generated by the maps

$$x_{\alpha_2}, \quad x_{\alpha_4}, \quad u \mapsto x_{\alpha_3}(u)\, x_{\alpha_5}(u), \quad u \mapsto x_{\alpha_1}(u)\, x_{\alpha_6}(u), \tag{9.13}$$

etc., where the displayed maps correspond to the roots α_1, α_2, α_3, and α_4 respectively in F_4, cf. [**Sp 98**, §10.3] or [**Dy 57b**, p. 194]. We claim that N is the direct product of F_4 with the center Z of E_6, which is isomorphic to $\boldsymbol{\mu}_3$.

Restricting the representation V of E_6 to F_4, we find that V is a direct sum of an F_4-invariant line $[v]$ (for some v) and an indecomposable 26-dimensional representation W (which is even irreducible if the characteristic of k_0 is not 3 [**GS**, p. 412]).

The maximal proper parabolics of L have Levi subgroups of type

$$D_5, \quad A_1 \times A_4, \quad A_1 \times A_2 \times A_2, \quad A_5,$$

and these semisimple parts all have dimension strictly less than 52, the dimension of F_4. Therefore F_4 is not contained in a proper parabolic subgroup of L. By [**Rö 93a**, Prop. 3.5], it follows that v belongs to the open L-orbit in V.

Clearly, F_4 is contained in the stabilizer N of $[v]$ in E_6, and by dimension count it is the identity component of N. Since Z is also contained in N, it suffices to prove that F_4 and Z generate the normalizer of F_4 in E_6. But every automorphism of F_4 is inner and F_4 has trivial center, so the normalizer of F_4 in E_6 is the product $F_4 \times C$, where C is the centralizer of F_4 in E_6. Therefore it suffices to prove that the center Z is all of C.

Write T_4 for the maximal torus $(F_4 \cap T)^\circ$ of F_4. The centralizer of T_4 in E_6 contains T, is reductive [**Bor**, 13.17, Cor. 2a], and is generated by T and the images of the x_γ's, where γ varies over the roots of E_6 whose inner product with $\alpha_2, \alpha_4, \alpha_3 + \alpha_5$, and $\alpha_1 + \alpha_6$ is zero. Such a root γ is a $\mathbb{Q}$-linear combination of the weights

$$\omega_3 - \omega_5 = \frac{1}{3}(\alpha_1 + 2\alpha_3 - 2\alpha_5 - \alpha_6) \quad \text{and} \quad \omega_1 - \omega_6 = \frac{1}{3}(2\alpha_1 + \alpha_3 - \alpha_5 - 2\alpha_6).$$

But such a γ would have disconnected support,[i] which is impossible by [**Bou Lie**, §VI.1.6, Cor. 3a to Prop. 19]. So the centralizer of T_4 in E_6 is T, and in particular C is contained in T. But C commutes with the images of the maps in (9.13), hence with the image of x_{α_i} for $1 \le i \le 6$. That is, C is contained in the center Z of E_6. This completes the proof that N equals $F_4 \times Z$.

Combining this example with Th. 9.11 gives that every k-E_6-torsor can be written (not necessarily uniquely) as a pair (J, β), where J is an Albert k-algebra and β belongs to $k^\times / k^{\times 3}$. For a classical proof of this in characteristic $\ne 2, 3$, see [**Sp 62**]. For an application, see [**GH**, §5] or 11.9 below.

9.14. Transitivity over k_{sep}. Maintain the notation of the rest of this section, and write k_{alg} for an algebraic closure of k. The dense subset $\mathcal{O}(k_{\mathrm{alg}})$ of $\mathbb{P}(V)$ is the $G(k_{\mathrm{alg}})$-orbit of $[v]$. If the stabilizer N of $[v]$ is smooth, then $G(k_{\mathrm{sep}})$ *acts transitively on the k_{sep}-points of $\mathcal{O}(k_{\mathrm{alg}})$*. To see this, fix $[w] \in \mathbb{P}(V)(k_{\mathrm{sep}})$ and $g \in G(k_{\mathrm{alg}})$ such that $g \cdot [v]$ is $[w]$. The set of elements of G that sends $[v]$ to $[w]$—the coset gN—is defined over k_{sep}. As N is a variety, so is gN, and the k_{sep}-points are dense in gN by [**Sp 98**, 11.2.7], proving the claim.

Context. The representations appearing in 9.3 are nearly the same as the prehomogeneous vector spaces appearing in [**SK**]. Recall that a *prehomogeneous vector space* is a representation V of an algebraic group G such that there is an open G-orbit in V. These too lead to surjections in cohomology, by the same proof as in 9.3. However, we are interested in the case where G is semisimple (and not merely reductive), for which there are not enough prehomogeneous vector spaces.

Continuing the comparison of G-orbits in V and $\mathbb{P}(V)$, we note that in the examples of 9.11 considered below (listed in Table 23a), the G-orbit of v in the affine space V is a hypersurface, more specifically a level set of a homogeneous G-invariant polynomial on V. However, this need not be true, as considering $G = \mathrm{Spin}_{10}$ shows: Viewing G as a subgroup of E_6, the recipe of 9.11 gives that V is a half-spin representation, and the G-orbit of v in that case is dense in V [**Ig**, Prop. 2]. (In Example 17.8, we view G as a subgroup of Spin_{12}, the resulting V is the 10-dimensional vector representation, and the G-invariant polynomial on V is the quadratic form.)

In the setup for Th. 9.11, we cited [**ABS**] because it is a convenient reference, but the core idea can certainly be found in other, earlier references, e.g., [**Vi**].

10. New invariants from homogeneous forms

A *(homogeneous) form of degree d* on a k_0-vector space V is a nonzero element of the d-th symmeric power $S^d(V^*)$. Equivalently, fixing a basis $x_1, x_2, \ldots, x_n$ for

[i] Recall that every root γ can be written uniquely as an integral linear combination of the simple roots. The *support* of γ is the set of those simple roots whose coefficient is nonzero in this expression.

the dual space V^*, it is a homogeneous polynomial of degree d in $k_0[x_1, x_2, \ldots, x_n]$. In this section, we give a mechanism for constructing new invariants of a group G from G-invariant forms.

Suppose that V is a representation of an algebraic group G and that V supports a G-invariant form f. Each $y \in H^1(k, G)$ defines a twisted form f_y on $V \otimes k$.

We are concerned with the case where f is a form of degree d such that $dC = 0$. For an invariant $a \in \mathrm{Inv}_{k_0}(G, C)$ and $v \in V \otimes k$ such that $f_y(v)$ is not zero, we consider the element

$$a(y) \cdot (f_y(v)) \quad \in M(k, C). \tag{10.1}$$

(We view $(f_y(v))$ as an element of $H^1(k, \boldsymbol{\mu}_d)$.) The following proposition is adapted from [**Rost 99c**, Prop. 5.2].

10.2. PROPOSITION. *If $a(y)$ is zero whenever f_y has a nontrivial zero, then the element* (10.1) *depends only on y (and not on the choice of v) and the map $y \mapsto a(y) \cdot (f_y(v))$ defines an invariant of G over k_0.*

Recall that f_y is said to *have a nontrivial zero* if there is some nonzero $v \in V \otimes k$ such that $f_y(v) = 0$. (Obviously, f_y always has the "trivial" zero $f_y(0) = 0$.)

10.3. EXAMPLE. In the "smallest" case, when V is 1-dimensional, we can see the proposition directly. If we fix a dual basis x for V, then f is αx^d for some $\alpha \in k_0^\times$. The action of G on V is given by a homomorphism $\chi : G \to \boldsymbol{\mu}_d$ and this defines an invariant $\underline{\chi} : H^1(*, G) \to H^1(*, \boldsymbol{\mu}_d)$. For $y \in H^1(k, G)$, f_y is the form $\alpha \chi(y)^{-1} x^d$, and for nonzero $v \in V \otimes k$, we have $(f_y(v)) = (\alpha) - (\chi(y))$. In particular, (10.1) is the value of the invariant $(\alpha) \cdot a - \underline{\chi} \cdot a$ at y.

PROOF OF PROPOSITION 10.2. By the example, we may assume the dimension of V is at least 2. Fix a basis for V^* as above. Writing f_y (viewed as an element of $k[V]$) in terms of this basis is equivalent to evaluating f_y at the generic point of V. Put

$$\omega := a(y) \cdot (f_y) \quad \in M(k(V), C).$$

We claim that ω is the restriction of some $\omega_0 \in M(k, C)$. By S10.1, it suffices to check that ω is unramified at every discrete valuation of $k(V)/k$ that corresponds to an irreducible hypersurface in V. Such a hypersurface is defined by some irreducible $\pi \in k[V]$. If π does not divide f_y (i.e., the hypersurface is not a component of the variety $\{f_y = 0\}$), then ω is unramified on the hypersurface by definition. If π does divide f_y in $k[V]$, we write $f_y = \pi^n \varepsilon$ for some ε not divisible by π, so

$$\omega = a(y) \cdot (\varepsilon) + n\, a(y) \cdot (\pi).$$

The residue of the first term is zero and the residue of the second is a multiple of $\mathrm{res}_{k(\pi)/k}\, a(y)$. The form f_y is zero on the sum of the vectors in the dual basis in $V \otimes k(\pi)$, and this is a nontrivial zero because the dimension of V is not 1. It follows that ω has residue zero. This proves the claim.

Specializing the generic point to $v \in V \otimes k$ maps $f_y \mapsto f_y(v)$ and $\omega \mapsto (a(y)) \cdot (f_y(v))$, but does not change ω_0. This proves that $a(y) \cdot (f_y(v))$ does not depend on the choice of v. The remainder of the proposition is clear. $\square$

10.4. EXAMPLE. The invariants produced by the proposition need not be interesting.

(1) Suppose that—in the situation of the proposition—the form f_y represents 1 for every $y \in H^1(k, G)$ and every k/k_0. (For example, take G to be the split group of type G_2 and let f_y be the norm on the octonion algebra corresponding to y.) Then the invariant given by the proposition is identically zero.

(2) Suppose that a is an invariant as in the proposition and that k_0 contains a primitive d-th root of unity. (This second hypothesis holds if f is a quadratic form.) Applying the proposition once produces an invariant a', and this new invariant also satisfies the hypothesis of the proposition. Applying the proposition again, we obtain an invariant

$$a'' \colon y \mapsto a(y) \cdot (f_y(v)) \cdot (f_y(v)) = -a(y) \cdot (f_y(v)) \cdot (-1),$$

where the second equality is by [**Dr**, p. 82, Cor. 5]. That is, a'' equals $-a(y) \cdot (-1)$.

11. Mod 3 invariants of simply connected E_6

In this section we assume that the characteristic of k_0 is different from 2 and 3.

11.1. INVARIANTS OF THE SPLIT E_6. We compute the invariants of the simply connected split group of type E_6, which we denote simply by E_6. (The mod 3 invariants of the *adjoint* split E_6 are not known.) The mod 2 invariants were computed in Exercise 22.9 in S. We note that E_6 has no invariants modulo primes $\neq 2, 3$ by the remarks in 5.6.

As in Example 9.12, we have an inclusion

$$i \colon F_4 \times \boldsymbol{\mu}_3 \hookrightarrow E_6$$

that identifies $\boldsymbol{\mu}_3$ with the center of E_6 such that the induced map

$$i_* \colon H^1(*, F_4 \times \boldsymbol{\mu}_3) \to H^1(*, E_6) \tag{11.2}$$

is a surjection. Two classes (J, β) and (J', β') have the same image in $H^1(k, E_6)$ if and only if there is a vector space isomorphism $f \colon J \to J'$ such that $\beta N_J = \beta' N_{J'} f$, where N_J and $N_{J'}$ denote the cubic norms on J and J', see [**Ga 01b**, 2.8(2)].

11.3. EXERCISE. Albert algebras J, J' are isotopic (see [**J 68**] for a definition) if and only if their norm forms are similar, i.e., $i_*(J, \beta) = i_*(J', \beta')$ for some $\beta, \beta' \in k^\times$, see pages 242–244 of [**J 68**]. Prove that J and J' are isotopic if and only if their norms are *isomorphic*, i.e., $i_*(J, 1) = i_*(J', 1)$. Prove also that $i_*(J, 1) = i_*(J, \beta)$ if and only if β is the norm of an element of J.

Composing (11.2) and (8.3) gives a functor

$$H^1(*, (\mathrm{PGL}_3 \times \boldsymbol{\mu}_3) \times \boldsymbol{\mu}_3) \to H^1(*, E_6)$$

where the $\mathrm{PGL}_3 \times \boldsymbol{\mu}_3$ in parentheses is the subgroup of F_4 from §8. This functor is surjective at 3 because every Albert algebra is in the image of (8.3) after an extension of the base field of dimension at most 2. Therefore the restriction map

$$\mathrm{Inv}^{\mathrm{norm}}(E_6, \mathbb{Z}/3\mathbb{Z}) \to \mathrm{Inv}^{\mathrm{norm}}(\mathrm{PGL}_3 \times \boldsymbol{\mu}_3 \times \boldsymbol{\mu}_3, \mathbb{Z}/3\mathbb{Z}) \tag{11.4}$$

is injective.

11.5. AN INVARIANT OF DEGREE 3. Consider the invariant g_3 of $\mathrm{PGL}_3 \times \boldsymbol{\mu}_3 \times \boldsymbol{\mu}_3$ defined by

$$g_3 \colon (A, \alpha, \beta) \mapsto [A] \cdot (\alpha) \quad \in H^3(k, \boldsymbol{\mu}_3^{\otimes 2}) \tag{11.6}$$

for (A, α, β) defined over k. We now give two arguments to show that it is in the image of (11.4).

PROOF #1. If (A, α, β) and (A', α', β') have the same image in $H^1(k, E_6)$, then the Albert algebras $J(A, \alpha), J(A', \alpha')$ have similar norms. But as they are first Tits constructions, this implies the algebras are isomorphic [**PeR 84**, 4.9], hence $[A]\cdot(\alpha)$ equals $[A']\cdot(\alpha')$ as in 8.5. That is, g_3 satisfies (7.2) and so extends uniquely to an invariant of E_6. □

PROOF #2. The Dynkin index of E_6 is 6 [**Mer**, 16.6], so the mod 3 portion of the Rost invariant defines a nonzero invariant

$$g_3' \colon H^1(*, E_6) \to H^3(*, \boldsymbol{\mu}_3^{\otimes 2}).$$

As the inclusion $F_4 \hookrightarrow E_6$ has Rost multiplier 1 [**Ga 01a**, 2.4], the restriction of g_3' to $H^1(*, F_4)$ is εg_3 for $\varepsilon = \pm 1$ and g_3 the invariant from 8.5. The composition

$$H^1(k, F_4) \times H^1(k, \boldsymbol{\mu}_3) \to H^1(k, E_6) \xrightarrow{g_3'} H^3(k, \boldsymbol{\mu}_3^{\otimes 2})$$

sends an Albert k-algebra J and a $\beta \in k^\times/k^{\times 3}$ to the element $\varepsilon g_3(J)$. (The case $\beta = 1$ is already done. In general, one uses a twisting argument as in [**GQ a**, Remark 2.5(i)].) The invariant $\varepsilon g_3'$ restricts to the map g_3 from (11.6). □

We abuse notation and write also g_3 for the invariant of E_6 that restricts to the g_3 from (11.6). Note that the image of this invariant of E_6 consists of symbols in $H^3(k, \boldsymbol{\mu}_3^{\otimes 2})$, because the same is true for the invariant g_3 of F_4.

11.7. AN INVARIANT OF DEGREE 4. Define an invariant g_4 of $\mathrm{PGL}_3 \times \boldsymbol{\mu}_3 \times \boldsymbol{\mu}_3$ by putting

$$g_4 \colon (A, \alpha, \beta) \mapsto [A]\cdot(\alpha)\cdot(\beta) \quad \in H^4(k, \boldsymbol{\mu}_3^{\otimes 3}). \tag{11.8}$$

We give two proofs of the fact that g_4 extends to an invariant $H^1(*, E_6) \to H^4(*, \boldsymbol{\mu}_3^{\otimes 3})$.

PROOF #1. We check (7.2). Suppose that (A, α, β) and (A', α', β') have the same image in $H^1(k, E_6)$. As in 11.5, $J(A, \alpha)$ and $J(A', \alpha')$ are isomorphic and $[A]\cdot(\alpha)$ equals $[A']\cdot(\alpha')$. Further, β/β' is a similarity of the norm of $J(A, \alpha)$. By Exercise 11.3, β/β' is a norm from $J(A, \alpha)$, hence $J(A, \alpha)$ is isomorphic to $J(A'', \alpha'')$ for some central simple algebra A'' such that β/β' is reduced norm from A'' [**PeR 84**, 4.2]. We conclude that

$$[A]\cdot(\alpha)\cdot(\beta) - [A']\cdot(\alpha')\cdot(\beta') = [A]\cdot(\alpha)\cdot(\beta/\beta') = [A'']\cdot(\alpha'')\cdot(\beta/\beta') = 0.$$

This verifies (7.2), hence g_4 extends to an invariant of $H^1(*, E_6)$. □

PROOF #2 (SKETCH). Observe that $H^1(k, E_6)$ classifies cubic forms that become isomorphic to the norm of an Albert algebra over a separable closure of k. The statement "i_* is surjective" says that such a cubic form is a scalar multiple—say, $\beta.N_J$—of the norm on an Albert k-algebra J. Moreover, $g_3(J)$ is zero whenever the norm N_J has a nontrivial zero (i.e., whenever J is reduced), so Prop. 10.2 gives that the map

$$g_4 \colon \beta.N_J \mapsto g_3(J)\cdot(\beta)$$

is a well-defined invariant of E_6. □

As usual, we write also g_4 for the invariant of E_6 that restricts to give the g_4 defined in (11.8).

11.9. PROPOSITION. $\mathrm{Inv}^{\mathrm{norm}}_{k_0}(E_6, \mathbb{Z}/3\mathbb{Z})$ *is a free* $R_3(k_0)$*-module with basis* g_3, g_4.

PROOF. We imitate the proofs of Propositions 6.1 and 8.6. The restriction map

$$i^* : \mathrm{Inv}^{\mathrm{norm}}_{k_0}(E_6, \mathbb{Z}/3\mathbb{Z}) \to \mathrm{Inv}^{\mathrm{norm}}_{k_0}(F_4 \times \boldsymbol{\mu}_3, \mathbb{Z}/3\mathbb{Z})$$

is an injection.

The center of E_6 is contained in a maximal split torus, hence the image of the map $H^1(*, \boldsymbol{\mu}_3) \to H^1(*, E_6)$ is zero. Applying 6.7 and Propositions 2.1 and 8.6, we find that g_3 and g_4 span $\mathrm{Inv}^{\mathrm{norm}}_{k_0}(E_6, \mathbb{Z}/3\mathbb{Z})$. □

11.10. EXERCISE (Mod 3 invariants of the quasi-split E_6). For K a quadratic field extension of k_0, write E_6^K for the simply connected quasi-split group of type E_6 associated with the extension K/k. Describe the "mod 3" invariants of E_6^K.

11.11. OPEN PROBLEM. [**PeR 94**, p. 205, Q. 4] Let J, J' be Albert k-algebras. If J and J' have similar norms, then their images in $H^1(k, E_6)$ are the same, hence they have the same Rost invariant. In the notation of §8 of this note and §22 of S, $f_3(J) = f_3(J')$ and $g_3(J) = g_3(J')$. Does the converse hold? That is, if $f_3(J) = f_3(J')$ and $g_3(J) = g_3(J')$, are the norms of J and J' necessarily similar?

[If J and J' are reduced, the answer is "yes", see [**J 68**, p. 369, Th. 2].]

12. Surjectivities: the highest root

We now describe a general situation where — in the setting of 9.11 — we can describe the identity component N° of the stabilizer. We will use this to apply Th. 9.11 to the simply connected group of type E_7.

12.1. Let $\widetilde{G}$ be a simply connected split algebraic group *not* of type A. The highest root $\widetilde{\alpha}$ is connected to a unique simple root—see Table 9—which we take to be π in the notation of 9.11. This situation was studied by Röhrle in [**Rö 93b**], and for convenience of reference, we adopt the hypotheses of his main theorem. Namely, we assume that π is long (equivalently, $\widetilde{G}$ is not of type C), the rank of G is at least 4, and the characteristic is $\neq 2$.

As $-\widetilde{\alpha}$ is joined to π by a single bond, $\widetilde{\alpha}$ is the fundamental weight corresponding to the simple root π, i.e., for every root β, the integer $\langle \widetilde{\alpha}, \beta \rangle$ is the coordinate of π in β. For example, the π-coordinate of $\widetilde{\alpha}$ is $\langle \widetilde{\alpha}, \widetilde{\alpha} \rangle = 2$. For w_0 the longest element of the Weyl group of $\widetilde{G}$, clearly $w_0(\widetilde{\alpha}) = -\widetilde{\alpha}$, hence $w_0(\pi) = -\pi$.

We take V to be $Q(1)/Q(2)$, where Q is the unipotent radical of the parabolic subgroup opposite to the one chosen in 9.11, so that Q is generated over k_{sep} by the U_α where α has positive π-coordinate. We do this both to agree with Röhrle's notation and for the convenience of working with positive roots. As mentioned in Remark 9.10.1, this V is the dual of the irreducible L-module with highest weight $\widetilde{\alpha}$, meaning it has highest weight $-w_0(\widetilde{\alpha}) = \widetilde{\alpha}$ also. Changing from the parabolic in 9.11 to its opposite has not changed the isomorphism class of V.

Table 12 describes the possibilities we consider. The last two columns will be explained below. Readers who know some nonassociative algebra will immediately recognize that V must be a Freudenthal triple system. Some convenient comparisons are [**Mey**, (8.4)] and [**Kr**, Table 1].

The top five rows of the table are "sisters": The groups $\widetilde{G}$ from these rows form the bottom row of Freudenthal's "magic square", resp. the G's form the next-to-the-bottom row. The representations V are the "preferred representations" from

$\widetilde{G}$	G	V	$\dim V$	$(N^\circ)^{\mathrm{ss}}$	$\dim Z(N^\circ)$
D_4	$(\mathrm{SL}_2)^{\times 3}$	$k^2 \otimes k^2 \otimes k^2$	8	1	2
F_4	Sp_6		14	SL_3	0
E_6	SL_6	$\wedge^3 k^6$	20	$\mathrm{SL}_3 \times \mathrm{SL}_3$	0
E_7	Spin_{12}	half-spin	32	SL_6	0
E_8	E_7	minuscule	56	E_6	0
Spin_d $(d \geq 9)$	$\mathrm{SL}_2 \times \mathrm{Spin}_{d-4}$	$k^2 \otimes$ vector	$2d-8$	Spin_{d-6}	1

TABLE 12. Internal Chevalley modules corresponding to the highest root

The last line of the table combines the cases where $\widetilde{G}$ is of type B_n $(n \geq 4)$ or D_n $(n \geq 5)$.

the bottom row of the magic triangle in [**DG 02**, Table 2]. The top five rows of Table 12 are related to cubic Jordan algebras of dimension 3, 6, 9, 15, and 27 respectively. The little one (with $\dim V = 8$) has appeared in Bhargava's work on higher reciprocity laws [**Bh**, p. 220] and in the solution to the Kneser-Tits Problem for rank 1 groups of type ${}^3D_4$ and ${}^6D_4$ from [**Pr**]. Note that in [**Pr**], the group G is not split, but π is circled in the index of G. In that case, the representation V is defined and there is an open G-orbit in $\mathbb{P}(V)$ as in 9.11. The stabilizer of a k-point in the open orbit is a "twisted form" of the N we now study.

Write $\widetilde{\Phi}$ for the set of roots of $\widetilde{G}$. We view $\widetilde{G}$ as defined by generators and relation as in [**St**]. In particular, for each root $\alpha \in \widetilde{\Phi}$, the unipotent subgroup U_α is the image of a homomorphism $x_\alpha \colon \mathbb{G}_a \to G$.

We put

$$v := x_\pi(r) x_{\widetilde{\alpha}-\pi}(s) U_{\widetilde{\alpha}} \quad \in V$$

for some $r, s \in k^\times$. It belongs to the open L-orbit in V [**Rö 93b**, 4.4], and we define N to be the scheme-theoretic stabilizer of $[v]$ in G.

12.2. LEMMA ($\operatorname{char} k_0 \neq 2$). *The identity component N° of N is reductive. Its semisimple part is simply connected and generated by the subgroups U_β as β varies over the roots in $\widetilde{\Phi}$ whose support contains neither π nor any root adjacent to π. The rank of its central torus equals* $\deg \pi - 1$.

The notation $\deg \pi$ denotes the degree of the vertex π of the Dynkin diagram, i.e., the number of simple roots that are distinct from and not orthogonal to π. The lemma says that the Dynkin diagram of N is obtained from the Dynkin diagram of $\widetilde{G}$ by deleting π and every vertex adjacent to π.

PROOF. Let $\beta \in \widetilde{\Phi}$ be as in the statement of the lemma. The support of $\pi \pm \beta$ has two connected components—the support of π and β—so $\pi \pm \beta$ is not a root of $\widetilde{G}$. For sake of contradiction, suppose that $\widetilde{\alpha} - \pi \pm \beta$ is a root of $\widetilde{G}$. It has π-coordinate 1, hence

$$\langle \widetilde{\alpha}, \widetilde{\alpha} - \pi \pm \beta \rangle = 1 \quad \text{and} \quad s_{\widetilde{\alpha}-\pi\pm\beta}(\widetilde{\alpha}) = \pi \mp \beta.$$

(Here and below we write s_β for the reflection defined by a root β.) That is, $\pi \mp \beta$ is a root of $\widetilde{G}$, a contradiction.

It follows from the previous paragraph that the subgroup H of G generated by the U_β's fixes v and so is a subgroup of N. The type of H is listed in the next-to-the-last column of Table 12, and H is simply connected by [**SS**, 5.4b]. We note that, line-by-line in the table, H has dimension

$$0, 8, 16, 35, 78, \frac{d^2 - 13d}{2} + 21.$$

Next consider the largest subtorus T_Z of $\widetilde{T}$ on which π, $\widetilde{\alpha}$, and the simple roots belonging to H vanish. This torus belongs to N, commutes with H, and has dimension

$$\operatorname{rank} \widetilde{G} - \operatorname{rank} H - 2 = \deg \pi - 1.$$

This number is listed in the last column of Table 12.

The subgroup $H.T_Z$ of G is connected and reductive with derived subgroup H. To complete the proof of the lemma, it suffices to check that $H.T_Z$ and N have the same dimension, i.e., to to check the equation

$$\dim H + \dim T_Z = \dim G - \dim V + 1. \tag{12.3}$$

The dimension of G, line-by-line in the table, is

$$9, 21, 35, 66, 133, \frac{d^2 - 9d}{2} + 13,$$

so equation (12.3) holds in each case. □

12.4. Orthogonal long roots in $\widetilde{\Phi}_1$. We put $\widetilde{\Phi}_j$ for the roots whose π-coordinate is j. For a positive root $\beta \in \widetilde{\Phi}$, the π-coordinate of β is 0, 1, or 2, and it is 2 if and only if β equals $\widetilde{\alpha}$, see [**Bou Lie**, §VI.1.8, Prop. 25(iv)]. That is, $\widetilde{\Phi}_j$ is nonempty only for $j = 0, \pm 1, \pm 2$ and $\widetilde{\Phi}_2$ is the singleton $\{\widetilde{\alpha}\}$.

As in [**Rö 93b**, p. 145], there is a sequence $\mu_1, \mu_2, \mu_3, \mu_4$ of pairwise orthogonal long roots in $\widetilde{\Phi}_1$.

Lemma. *The roots $\mu_1, \mu_2, \mu_3, \mu_4$ are pairwise strongly orthogonal.*

Recall that roots β, γ are said to be *strongly orthogonal* if $\beta + \gamma$ and $\beta - \gamma$ are not roots and $\beta \neq \pm\gamma$.

Proof. If $\mu_i + \mu_j$ is a root, then it has π-coordinate 2, hence it equals $\widetilde{\alpha}$. But

$$0 = \langle \mu_i, \mu_j \rangle = \langle \widetilde{\alpha} - \mu_j, \mu_j \rangle = -1,$$

a contradiction. Further, μ_i and μ_j are orthogonal, so since $\mu_i + \mu_j$ is not a root, neither is $\mu_i - \mu_j$. □

12.5. Strongly orthogonal roots in G. The Weyl group of G acts transitively on the roots in $\widetilde{\Phi}_1$ of the same length [**ABS**, §2, Lemma 1], so we may assume that μ_1 equals π. For $j = 2, 3, 4$, we set:

$$\gamma_j := \widetilde{\alpha} - \pi - \mu_j.$$

LEMMA. *$\gamma_2, \gamma_3, \gamma_4$ are pairwise strongly orthogonal long roots of G. For various x, y, the value of $\langle x, y\rangle$ is given by the table:*

		y: $\widetilde{\alpha}$	π	μ_j	γ_j
	$\widetilde{\alpha}$	2	1	1	0
x	π	1	2	0	-1
	μ_j	1	0	2	-1
	γ_j	0	-1	-1	2

PROOF. The top row of the table is the π-coordinate of y, and we know these already. We calculate that γ_j equals $s_\pi s_{\mu_j}(\widetilde{\alpha})$, so in particular, γ_j has the same length as $\widetilde{\alpha}$: long. As all the roots in table have the same length, the table is symmetric. As for π and μ_j, they are orthogonal by construction. The entries for $\langle \gamma_j, \pi\rangle$ and $\langle \gamma_j, \mu_j\rangle$ are straightforward computations.

For $i \neq j$ we have $\langle \gamma_j, \mu_i\rangle = 1$, hence $\langle \gamma_i, \gamma_j\rangle = 0$. Further, $\gamma_i - \gamma_j = \mu_j - \mu_i$ is not a root by Lemma 12.4. As in the proof of that lemma, $\gamma_i + \gamma_j$ is not a root. $\square$

We note that $\mu_2 + \mu_3 + \mu_4 = 2\widetilde{\alpha} - \pi$ [**Rö 93b**, 1.4], so

$$\gamma_2 + \gamma_3 + \gamma_4 = \widetilde{\alpha} - 2\pi. \tag{12.6}$$

12.7. A COPY OF SL_2. Define 1-parameter subgroups $x, y\colon \mathbb{G}_a \to G$ via

$$x(u) := \prod_{j=2}^{4} x_{\gamma_j}(u) \quad \text{and} \quad y(u) := \prod_{j=2}^{4} x_{-\gamma_j(u)}.$$

Since the γ_j's are strongly orthogonal, the images of the x_{γ_j} commute [**St**, p. 30, (R2)], i.e., it does not matter in what order the displayed products are written. The images of x and y generate a copy of SL_2 in G that we denote simply by SL_2. For $t \in \mathbb{G}_m$, we set

$$w(t) := x(t)y(-t^{-1})x(t) \quad \text{and} \quad h(t) := w(t)w(-1).$$

The map h is a homomorphism and its image is a maximal torus in SL_2. How does SL_2 act on V? Identity (R8) from [**St**, p. 30] says that for roots β, δ, we have:

$$h_\beta(t)x_\delta(u) = x_\delta(t^{\langle\delta,\beta\rangle}u)h_\beta(t) \tag{R8}$$

where $h_\beta\colon \mathbb{G}_m \to \widetilde{T}$ is the cocharacter corresponding to the coroot $\check{\beta}$. In particular,

$$h(t)x_\pi(u) = x_\pi(t^{-3}u)h(t) \quad \text{and} \quad h(t)x_{\widetilde{\alpha}-\pi}(u) = x_{\widetilde{\alpha}-\pi}(t^3u)h(t) \tag{12.8}$$

Moreover, we have:

LEMMA. *There exists a $c \in \{\pm 1\}$ such that*

$$w(t)x_\pi(u) = x_{\widetilde{\alpha}-\pi}(ct^3u)w(t) \quad \text{and} \quad w(t)x_{\widetilde{\alpha}-\pi}(u) = x_\pi(-ct^{-3}u)w(t)$$

for all $t \in \mathbb{G}_m$ and $u \in \mathbb{G}_a$.

PROOF. Steinberg gives the formula [**St**, p. 67, Lemma 37a]:

$$w_\beta(t)x_\delta(u) = x_{s_\beta\delta}(c(\beta,\delta)t^{-\langle\delta,\beta\rangle}u)w_\beta(t),$$

where $w_\beta(t)$ is defined to be $x_\beta(t)x_{-\beta}(-t^{-1})x_\beta(t)$ and $c(\beta,\delta) = \pm 1$ depends only on β and δ. Applying this with $\delta = \pi$ and successively with $\beta = \gamma_2, \gamma_3, \gamma_4$, we find $c \in \{\pm 1\}$ such that

$$w(t)x_\pi(u) = x_{\widetilde{\alpha}-\pi}(ct^3u)w(t).$$

(For the exponent of t, note e.g. that $\langle s_{\gamma_2}\pi, \gamma_3\rangle = \langle \pi, s_{\gamma_2}\gamma_3\rangle = \langle \pi, \gamma_3\rangle = -1$.) Similarly, we obtain

$$w(t)x_{\widetilde{\alpha}-\pi}(u) = x_\pi(c't^{-3}u)w(t)$$

for some $c' \in \{\pm 1\}$.

The equations (12.8) give $h(-1)x_\pi(u) = x_\pi(-u)h(-1)$ and since $h(-1) = w(-1)^2$, we have:

$$x_\pi(-u)h(-1) = w(-1)^2 x_\pi(u) = x_\pi(cc'u)h(-1).$$

So $c' = -c$. □

12.9. Remark. We can describe this copy of SL_2 concretely in the notation of Dynkin [**Dy 57b**, Ch. III]. For simplicity, we consider the cases where $\widetilde{G}$ is simply laced, so we may identify roots and coroots by defining all roots to have length 2 with respect to the Weyl-invariant inner product $(\ ,\ )$. By (12.6), the intersection of the maximal torus $\widetilde{T}$ of $\widetilde{G}$ with SL_2 is the image of the cocharacter $h_{\widetilde{\alpha}-2\pi}$. For δ a simple root of G, the inner product $(\widetilde{\alpha} - 2\pi, \delta)$ is 2 if δ is adjacent to π and 0 otherwise. (Recall that π is not a root of G.) That is, Dynkin would denote the corresponding copy of $\mathfrak{sl}_2$ in the Lie algebra of G by attaching a 2 to the vertices of the Dynkin diagram of G that are adjacent to π.

12.10. We take N to be the scheme-theoretic stabilizer of

$$v := x_\pi(1)\, x_{\widetilde{\alpha}-\pi}(-c)\, U_{\widetilde{\alpha}} \quad \in V$$

for c as in Lemma 12.7. For a primitive 4-th root of unity i, we have $w(i)v = iv$. (See Remark 9.10 for the vector space structure on V.) The map $i \mapsto w(i)$ defines an injection $\boldsymbol{\mu}_4 \hookrightarrow N$, and we abuse notation by writing also $\boldsymbol{\mu}_4$ for the image in N.

So far, what we have written holds for the general setting of 12.1. We now specialize to the case where G is E_7. Write C for the rank 1 torus appearing in 12.9, i.e., for the image of the cocharacter $h_{2\omega_7} \colon \mathbb{G}_m \to \widetilde{T}$, which maps

$$t \mapsto h_{\alpha_1}(t^2)\, h_{\alpha_2}(t^3)\, h_{\alpha_3}(t^4)\, h_{\alpha_4}(t^6)\, h_{\alpha_5}(t^5)\, h_{\alpha_6}(t^4)\, h_{\alpha_7}(t^3).$$

12.11. Lemma. *The centralizer of E_6 in E_7 is the rank 1 torus C. The normalizer of E_6 in E_7 is the group generated by C, E_6, and the copy of $\boldsymbol{\mu}_4$ from 12.10.*

Proof. Write T_6 and T_7 for the maximal tori in E_6 and E_7 respectively, obtained by intersecting with the maximal torus $\widetilde{T}$ of E_8. We argue along the lines of Example 9.12. First note that the centralizer of T_6 in E_7 contains T_7, is reductive, and is generated by T_7 and root subgroups U_γ of E_7 for roots γ of E_7 whose inner product with the simple root α_i is zero for $1 \leq i \leq 6$. Such a γ is a multiple of the fundamental weight ω_7 with integer coefficients, i.e., an integer multiple of $2\omega_7$. However, $2\omega_7$ has height 27 and the highest root of E_7 has height 17, so no such γ exists. Therefore the centralizer of T_6 in E_7 is T_7.

It follows that the centralizer of E_6 in E_7 is the subgroup of T_7 formed by intersecting the kernels of the roots of E_6. This is a computation in terms of root systems: the character group of this centralizer is the quotient of the E_7 weight lattice by the sublattice generated by the α_i for $1 \leq i \leq 6$; this quotient is free of rank 1. Therefore the centralizer is a rank 1 torus in T_7. To prove the first claim in the lemma, it suffices to observe that the inner product $(\widetilde{\alpha} - 2\pi, \delta)$ is zero for every root δ of E_6, which is clear because $\widetilde{\alpha} - 2\pi$ equals $2\omega_7$.

The quotient group of "outer automorphisms" (automorphisms modulo inner automorphisms) of E_6 is $\mathbb{Z}/2\mathbb{Z}$, so to prove the claim about the normalizer it suffices to show that conjugation by a generator $w(i)$ of $\boldsymbol{\mu}_4(k_{\mathrm{sep}})$ gives an outer automorphism of E_6. As $\boldsymbol{\mu}_4$ belongs to N, it normalizes the identity component E_6 of N. Further, conjugation by $w(i)$ inverts elements of the maximal torus C of SL_2. But C contains the center of E_6 by the previous paragraph, so conjugation by $w(i)$ is an outer automorphism of E_6. □

12.12. REMARK. From Lemma 12.11, we see that C contains the centers of E_6 and E_7. Restricting the cocharacter $h_{2\omega_7}$ to $\boldsymbol{\mu}_3$ and $\boldsymbol{\mu}_2$ respectively, we find

$$\zeta \mapsto h_{\alpha_1}(\zeta^2)\, h_{\alpha_3}(\zeta)\, h_{\alpha_5}(\zeta^2)\, h_{\alpha_6}(\zeta) \quad \text{and} \quad \varepsilon \mapsto h_{\alpha_2}(\varepsilon)\, h_{\alpha_5}(\varepsilon)\, h_{\alpha_7}(\varepsilon).$$

The images of these maps are the centers of E_6 and E_7 respectively, cf. [**GQ a**, 8.2, 8.1].

12.13. EXAMPLE ($E_6 \rtimes \boldsymbol{\mu}_4 \subset E_7$). We now show that the inclusion $E_6 \rtimes \boldsymbol{\mu}_4 \subset E_7$ from 12.10 induces a surjection

$$H^1(k, E_6 \rtimes \boldsymbol{\mu}_4) \to H^1(k, E_7) \tag{12.14}$$

for every extension k/k_0. By Th. 9.11, it suffices to show that $E_6 \rtimes \boldsymbol{\mu}_4$ is the stabilizer N of $[v] \in \mathbb{P}(V)$ for v as in 12.10. The subgroup of the torus C stabilizing $[v]$ is the image of $\boldsymbol{\mu}_6$ by (12.8), which is the subgroup of C generated by the center of E_6 and the copy of $\boldsymbol{\mu}_2$ in $\boldsymbol{\mu}_4$. Combining Lemma 12.11 and the fact that E_6 and $\boldsymbol{\mu}_4$ belong to N, we conclude that N equals $E_6 \rtimes \boldsymbol{\mu}_4$.

The surjectivity of (12.14) can be interpreted as a statement about Freudenthal triple systems; see [**Ga 01b**, 4.15] for a precise statement and an algebraic proof.

13. Mod 3 invariants of E_7

The goal of this section is to compute the invariants of a split group of type E_7 (simply connected or adjoint) with values in $\mathbb{Z}/3\mathbb{Z}$. We write E_7 for the simply connected split group of that type, and we assume throughout this section that the characteristic of k_0 is $\neq 2, 3$. (Roughly speaking, we avoid characteristic 2 in order to use the results of the previous section, and we avoid characteristic 3 because we wish to describe the invariants mod 3, cf. Remark 2.4.) The "heavy lifting" was already done in the previous section.

Recall that the split group F_4 of that type can be viewed as a subgroup of E_6 as in Example 9.12.

13.1. LEMMA. *The inclusion $F_4 \subset E_7$ gives a morphism*

$$H^1(*, F_4) \to H^1(*, E_7)$$

that is surjective at 3.

PROOF. We have inclusions

$$F_4 \times \boldsymbol{\mu}_3 \subset E_6 \subset E_6 \rtimes \boldsymbol{\mu}_4 \subset E_7.$$

The first and third of these induce surjections on H^1 by Examples 9.12 and 12.13. The second inclusion gives a morphism that is surjective at 3.

The image of $\boldsymbol{\mu}_3$ in E_7 is the center of E_6 as in Remark 12.12, so the inclusion $F_4 \times \boldsymbol{\mu}_3 \subset E_7$ factors through the subgroup $F_4 \times C$. As $H^1(k, C)$ is zero for every k/k_0, the images of $H^1(k, F_4)$ and $H^1(k, F_4 \times \boldsymbol{\mu}_3)$ in $H^1(k, E_7)$ agree. The claim follows. □

13.2. MOD 3 INVARIANTS OF E_7. We now give two proofs that the invariant g_3 of F_4 defined in 8.5 extends to an invariant of E_7, which we will also denote by g_3.

PROOF #1. Let $J, J' \in H^1(k, F_4)$ be Albert algebras whose images in $H^1(k, E_7)$ agree. Then J and J' have similar norms by [**Fe**, 6.8] and $g_3(J)$ equals $g_3(J')$ by 11.5. Lemma 7.1 gives that g_3 extends to an invariant of E_7. □

PROOF #2. The Rost invariant of E_7 has order 12 [**Mer**, 16.7], so the mod 3 portion defines a nonzero invariant of E_7 with values in $\mathbb{Z}/3\mathbb{Z}$. Moreover, the inclusion $F_4 \subset E_7$ has Rost multiplier 1, so the restriction of this invariant of E_7 to F_4 is — up to sign — the g_3 from 8.5. □

Combining Lemmas 5.3 and 13.1, we conclude that the restriction

$$\mathrm{Inv}_{k_0}^{\mathrm{norm}}(E_7, \mathbb{Z}/3\mathbb{Z}) \to \mathrm{Inv}_{k_0}^{\mathrm{norm}}(F_4, \mathbb{Z}/3\mathbb{Z})$$

is injective; by the above and Th. 8.6, it is an isomorphism. We conclude:

THEOREM. *$\mathrm{Inv}_{k_0}^{\mathrm{norm}}(E_7, \mathbb{Z}/3\mathbb{Z})$ is a free $R_3(k_0)$-module with basis g_3.* □

13.3. EXERCISE (Mod 3 invariants of adjoint E_7). Write E_7^{adj} for the split adjoint group of type E_7. Prove that the invariant g_3 of E_7 induces an invariant $g_3^{\mathrm{adj}}: H^1(*, E_7^{\mathrm{adj}}) \to H^3(*, \boldsymbol{\mu}_3^{\otimes 2})$ and that $\mathrm{Inv}_{k_0}^{\mathrm{norm}}(E_7^{\mathrm{adj}}, \mathbb{Z}/3\mathbb{Z})$ is a free $R_3(k_0)$-module with basis g_3^{adj}.

For the mod 2 invariants of E_7, the situation is much less clear.

13.4. OPEN PROBLEM. (Reichstein-Youssin [**RY**, p. 1047]) Let k_0 be an algebraically closed field of characteristic zero. Is there a nonzero invariant

$$H^1(*, E_7^{\mathrm{adj}}) \to H^8(*, \mathbb{Z}/2\mathbb{Z})?$$

[Some readers have expressed skepticism about the precise degree — 8 — suggested above. Nonetheless, the core of the question remains: What are the invariants of degree > 3?]

14. Construction of groups of type E_8

Write E_8 for the split algebraic group of that type over k_0. Since it is adjoint and every automorphism is inner, the set $H^1(k, E_8)$ is identified with the group of isomorphism classes of groups of type E_8 over k. (This same phenomenon occurs with groups of type G_2 and F_4.) Here we describe a construction of groups of type E_8 that is analogous to the first Tits construction of groups of type F_4 (equivalently, Albert algebras) from 8.1. The fruit of this construction will appear in the next section. As the results of §9 do not apply to E_8 by Example 9.7, we take a new approach here. We assume throughout this section that the characteristic of k_0 is $\neq 5$.

14.1. A SUBGROUP H OF E_8. Let G be split of type E_8. We write H for the subgroup of G generated by the root subgroups $U_{\pm\widetilde{\alpha}}$ and $U_{\pm\alpha_i}$ for $i \neq 5$. This subgroup is of type $A_4 \times A_4$. We identify the "first" component of H — generated by $U_{\pm\alpha_i}$ for $i = 1, 2, 3, 4$ — with SL_5 via an irreducible representation whose highest weight is 1 on α_1 and 0 on α_2, α_3, and α_4. We identify the "second" component of H with SL_5 via an irreducible representation whose highest weight is 1 on α_6 and 0 on α_7, α_8, and $-\widetilde{\alpha}$.

Write ϱ for the homomorphism $\boldsymbol{\mu}_5 \to E_8$ defined by

$$\varrho\colon \zeta \mapsto h_{\alpha_1}(\zeta) h_{\alpha_2}(\zeta^4) h_{\alpha_3}(\zeta^2) h_{\alpha_4}(\zeta^3). \tag{14.2}$$

Applying the method described in [**GQ a**, §8], one finds that the image of ϱ is the center of both copies of SL_5 in E_8. More precisely, the canonical identification of the center of SL_5 with $\boldsymbol{\mu}_5$ is the map ϱ for the first component of H and ϱ^3 for the second component of H. That is, we have identified H with the quotient of $\mathrm{SL}_5 \times \mathrm{SL}_5$ by the subgroup generated by (ζ, ζ^2) for $\zeta \in \boldsymbol{\mu}_5$.

For $i = 1, 2$, write $\pi_i\colon H \to \mathrm{PGL}_5$ for the projection on the i-th factor.

14.3. LEMMA. *For $\eta \in H^1(k, H)$, write A_i for the central simple k-algebra of degree 5 defined by $\pi_i(\eta)$. Then $2[A_1] = [A_2]$ in the Brauer group of k.*

The twisted group H_η is isomorphic to $(\mathrm{SL}(A_1) \times \mathrm{SL}(A_2))/\boldsymbol{\mu}_5$.

PROOF. Consider the diagram with exact rows

$$\begin{array}{ccccccccc}
1 & \longrightarrow & \boldsymbol{\mu}_5 \times \boldsymbol{\mu}_5 & \longrightarrow & \mathrm{SL}_5 \times \mathrm{SL}_5 & \longrightarrow & \mathrm{PGL}_5 \times \mathrm{PGL}_5 & \longrightarrow & 1 \\
 & & {\scriptstyle q}\downarrow & & \downarrow & & \| & & \\
1 & \longrightarrow & \boldsymbol{\mu}_5 & \xrightarrow{\varrho} & H & \xrightarrow{\pi_1 \times \pi_2} & \mathrm{PGL}_5 \times \mathrm{PGL}_5 & \longrightarrow & 1
\end{array}$$

where q is given by $(x, y) \mapsto y/x^2$. The diagram commutes because $y/x^2 = x(y/x^3)$ and $(y/x^2)^3 = y(y/x^3)^2$. We obtain a commutative diagram with exact rows:

$$\begin{array}{ccccc}
1 & \longrightarrow & H^1(k, \mathrm{PGL}_5 \times \mathrm{PGL}_5) & \longrightarrow & \mathrm{Br}_5 \times \mathrm{Br}_5 \\
\downarrow & & \| & & \downarrow{\scriptstyle q} \\
H^1(k, H) & \xrightarrow{\pi_1 \times \pi_2} & H^1(k, \mathrm{PGL}_5 \times \mathrm{PGL}_5) & \xrightarrow{\delta} & \mathrm{Br}_5 .
\end{array}$$

In the Brauer group, we find the equation:

$$0 = \delta(\pi_1 \times \pi_2)(\eta) = q(\pi_1(\eta), \pi_2(\eta)) = -2[A_1] + [A_2]. \qquad \square$$

14.4. THE SUBGROUP C OF H. Write C for the group $\mathbb{Z}/5\mathbb{Z} \times \boldsymbol{\mu}_5$. We define a homomorphism $t\colon C \times \boldsymbol{\mu}_5 \to H$ such that t restricted to $\boldsymbol{\mu}_5$ is the map ϱ from (14.2) and the restriction of t to C is given by

$$t|_C(i, \zeta^j) = (v^i u^j, v^i u^{2j})$$

in the notation of 6.2, where ζ is a fixed primitive 5-th root of unity. The formula $uv = \zeta vu$ shows that t is indeed a group homomorphism.

For each extension k/k_0, there is an induced function

$$t_*\colon H^1(k, C \times \boldsymbol{\mu}_5) \to H^1(k, E_8). \tag{14.5}$$

Because $H^1(k, E_8)$ classifies groups of type E_8 over k, we view t_* as a *construction* of groups of type E_8 via Galois descent.

14.6. EXAMPLE. We now compute $t_*(\gamma, z)$ for various $\gamma \in H^1(k, C)$ and $z \in H^1(k, \boldsymbol{\mu}_5)$. Put $\eta := t_*(\gamma, 1)$ and write A_1 for $\pi_1(\eta)$ as in Lemma 14.3. Twisting SL_5 and E_8 by η, we find a subgroup $\mathrm{SL}(A_1)$ of $(E_8)_\eta$.

The diagram

$$\begin{array}{ccc} C & \xrightarrow{t} & H \\ & {\scriptstyle c}\searrow & \downarrow{\scriptstyle \pi_1} \\ & & \mathrm{PGL}_5 \end{array}$$

commutes for c as in 6.2, so $\pi_1(\eta) = c_*(\gamma)$.

(1) Suppose that $c_*(\gamma)$ is zero, i.e., A_1 is split. Then H_η is split by Lemma 14.3. Since η and $t_*(\gamma, z)$ — viewed as elements of $H^1(k, H)$ — differ by a central cocycle $t_*(1, z)$, the twisted group $H_{t_*(\gamma,z)}$ is also split. Hence E_8 twisted by $t_*(\gamma, z)$ is split. We conclude that $t_*(\gamma, z)$ is zero in $H^1(k, E_8)$.

(2) Suppose now that $z_1, z_2 \in H^1(k, \boldsymbol{\mu}_5)$ differ by a reduced norm from A_1. The inclusion ϱ of $\boldsymbol{\mu}_5$ into $H \subset E_8$ is unaffected by twisting by η, and we obtain a map $H^1(k, \boldsymbol{\mu}_5) \to H^1(k, (E_8)_\eta)$. The composition

$$H^1(k, \boldsymbol{\mu}_5) \xrightarrow{\ \varrho\ } H^1(k, (E_8)_\eta) \xrightarrow[\tau_\eta]{\cong} H^1(k, E_8),$$

where τ_η is the twisting isomorphism, sends z_i to $t_*(\gamma, z_i)$. However, the first arrow factors through $H^1(k, \mathrm{SL}(A_1))$, hence z_1 and z_2 have the same image in $H^1(k, E_8)$.

14.7. PROPOSITION. *The morphism*

$$t_* \colon H^1(*, C \times \boldsymbol{\mu}_5) \to H^1(*, E_8)$$

is surjective at 5.

PROOF. *Step 1.* We first show that — for H the subgroup of E_8 defined in 14.1 — the morphism $H^1(*, H) \to H^1(*, E_8)$ is surjective at 5.

Let ξ be in $Z^1(k, E_8)$. Fix a maximal and k-split torus T of E_8 and a k-defined maximal torus T' in the twisted group $(E_8)_\xi$. There is some $g \in E_8(k_{\mathrm{sep}})$ such that $g^{-1}T'g = T$, and replacing ξ with $\sigma \mapsto g^{-1}\xi_\sigma\, {}^\sigma g$, we may assume that $\xi_\sigma\sigma(T) = T$ for every $\sigma \in \mathrm{Gal}(k_{\mathrm{sep}}/k)$, i.e., that ξ takes values in $N_{E_8}(T)$.

The Galois group acts trivially on the Weyl group $N_{E_8}(T)/T$, so the image $\overline{\xi} \in Z^1(k, N_{E_8}(T)/T)$ is a continuous homomorphism $\overline{\xi} \colon \mathrm{Gal}(k_{\mathrm{sep}}/k) \to N_{E_8}(T)/T$. Fix a 5-Sylow subgroup S of $N_{E_8}(T)/T$. Take K to be the subfield of k_{sep} fixed by $\overline{\xi}^{-1}(S)$; it is an extension of K of dimension not divisible by 5.

Because S and a 5-Sylow in $N_H(T)/T$ both have order 5^2, there is a $\overline{w} \in (N_{E_8}(T)/T)(K)$ such that the image of the map $\sigma \mapsto \overline{w}^{-1}\overline{\xi}_\sigma\overline{w}$ is contained in $N_H(T)/T$. Further, T is K-split, so there is some $w \in N_{E_8}(T)(K)$ such that w maps to $\overline{w}$. Replacing ξ with $\sigma \mapsto w^{-1}\xi_\sigma\, {}^\sigma w$, we may assume that $\mathrm{res}_{K/k}(\xi) \in H^1(K, E_8)$ is in the image of $H^1(K, H)$.

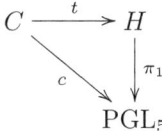

Step 2. We now show that the morphism $H^1(*, C \times \boldsymbol{\mu}_5) \to H^1(*, H)$ is surjective at 5. Fix $\eta \in Z^1(K, H)$ and let A be the central simple algebra of degree 5 representing $\pi_1(\eta) \in H^1(K, \mathrm{PGL}_5)$. By Lemma 6.6, there is an extension L/K of

dimension not divisible by 5 such that $A \otimes L$ is cyclic, i.e., equals $c_*(\gamma)$ for some $\gamma \in H^1(L, C)$. We have a commutative diagram with exact rows:

$$\begin{array}{ccccccccc} 1 & \longrightarrow & \boldsymbol{\mu}_5 & \longrightarrow & C \times \boldsymbol{\mu}_5 & \longrightarrow & C & \longrightarrow & 1 \\ & & \| & & \downarrow{\scriptstyle t} & & \downarrow & & \\ 1 & \longrightarrow & \boldsymbol{\mu}_5 & \xrightarrow{\varrho} & H & \xrightarrow{\pi_1 \times \pi_2} & \mathrm{PGL}_5 \times \mathrm{PGL}_5 & \longrightarrow & 1. \end{array}$$

By Lemma 14.3, γ and η have the same image in $H^1(k, \mathrm{PGL}_5 \times \mathrm{PGL}_5)$, namely the class of $(\pi_1(\eta), \pi_2(\eta))$. It follows that η and $t_*(\gamma)$ are in the same $H^1(k, \boldsymbol{\mu}_5)$-orbit. Fixing a $\lambda \in H^1(k, \boldsymbol{\mu}_5)$ such that $\eta = \lambda \cdot t_*(\gamma)$, we have:

$$t_*(\lambda \cdot \gamma) = \lambda \cdot t_*(\gamma) = \eta,$$

as desired. □

14.8. REMARK. The argument in Step 1 of the proof of the proposition amounts to the sort of argument behind [**BS**, Cor. 2.14] combined with Sylow's Theorem. It is essentially identical to the proof of Lemma 2 in [**Ch 95**], it is just written in a different language. This argument can be applied also in other situations, e.g., at the primes 3 and 2 with H replaced by $(E_6 \times \mathrm{SL}_3)/\boldsymbol{\mu}_3$ and a half-spin group of type D_8 respectively. This argument also appears for E_6 relative to the prime 3 in [**MPW**, p. 153].

14.9. THE ROST INVARIANT. We now compute the composition

$$H^1(k, C \times \boldsymbol{\mu}_5) \xrightarrow{t_*} H^1(k, E_8) \xrightarrow{r_{E_8}} H^3(k, \mathbb{Q}/\mathbb{Z}(2)) \tag{14.10}$$

for every extension k/k_0. As the Dynkin index of E_8 is $60 = 5 \cdot 12$ [**Mer**, 16.8], 5.4 says that the image of the composition is 5-torsion, hence lies in $H^3(k, \boldsymbol{\mu}_5^{\otimes 2})$. Recall that C is $\mathbb{Z}/5\mathbb{Z} \times \boldsymbol{\mu}_5$. We have:

14.11. LEMMA. *There is a uniquely determined $\lambda \in H^1(k_0, \boldsymbol{\mu}_5)$ and a natural number m not divisible by 5 such that the composition* (14.10) *is given by*

$$((x, y), z) \mapsto \lambda \cdot x \cdot y + m\, x \cdot y \cdot z$$

for every $x \in H^1(k, \mathbb{Z}/5\mathbb{Z})$ and $y, z \in H^1(k, \boldsymbol{\mu}_5)$ and every k/k_0.

PROOF. We first prove the claim in the case where z is zero and k_0 contains a primitive 5-th root of unity, which we use to identify $\boldsymbol{\mu}_5$ with $\mathbb{Z}/5\mathbb{Z}$. If x or y is zero, the class $t_*(x, y, 1)$ is zero in $H^1(k, E_8)$ by Example 14.6.1. Applying Lemma 6.7, we conclude that the composition (14.10) is $(x, y, 1) \mapsto \lambda \cdot x \cdot y$ for a unique $\lambda \in R_5(k_0)$. That is, the claim holds in this case.

We now consider the case where z is zero, but the extension k_1 obtained by adjoining a primitive 5-th root of unity to k_0 may be proper. By the previous paragraph, the restriction of (14.10) to $H^1(*, C)$ and viewed as an invariant $\mathsf{Fields}_{/k_1} \to \mathsf{Abelian\ Groups}$ is given by

$$(x, y) \mapsto \lambda_1 \cdot x \cdot y$$

for a uniquely determined $\lambda_1 \in H^1(k_1, \boldsymbol{\mu}_5)$. Write λ_0 for the unique class in $H^1(k_0, \boldsymbol{\mu}_5)$ whose restriction to k_1 is λ_1. Since the invariants (14.10) and $(x, y) \mapsto \lambda_0 \cdot x \cdot y$ agree over every extension k/k_1, Lemma 3.2 proves the claim.

Finally, we consider the general case. Put $\eta := t_*(x, y, 1)$ and consider the diagram

$$\begin{array}{ccccccc} H^1(k,\boldsymbol{\mu}_5) & \xrightarrow{\varrho} & H^1(k,\mathrm{SL}(A_1)) & \longrightarrow & H^1(k,(E_8)_\eta) & \xrightarrow{r_{(E_8)_\eta}} & H^3(k,\mathbb{Q}/\mathbb{Z}(2)) \\ & \searrow^{t_*(x,y,?)} & & & \cong \downarrow \tau_\eta & & \downarrow ?+r_{E_8}(\eta) \\ & & & & H^1(k,E_8) & \xrightarrow{r_{E_8}} & H^3(k,\mathbb{Q}/\mathbb{Z}(2)) \end{array}$$

where τ_η is the twisting isomorphism. The triangle obviously commutes and the square commutes by [**Gi 00**, p. 76, Lemma 7]. The image of $z \in H^1(k, \boldsymbol{\mu}_5)$ in the lower right corner going counterclockwise is $r_{E_8}(t_*(x, y, z))$, i.e., the image of (x, y, z) under (14.10). Since the inclusion of $\mathrm{SL}(A_1)$ in $(E_8)_\eta$ has Rost multiplier 1, the composition on the top row is $z \mapsto m\, x \cdot y \cdot z$ for some natural number m not divisible by 5. This proves the claim. □

14.12. A TWISTED MORPHISM. Fix a 1-cocycle $\mu \in Z^1(k, \boldsymbol{\mu}_5)$ such that $\mu = -m^*\lambda$ in $H^1(k, \boldsymbol{\mu}_5)$, where m^* denotes a natural number such that mm^* is congruent to 1 mod 5, and λ is as in Lemma 14.11. Define i to be the composition

$$i\colon H^1(*, C \times \boldsymbol{\mu}_5) \xrightarrow{\mu+?} H^1(*, C \times \boldsymbol{\mu}_5) \xrightarrow{t_*} H^1(*, E_8).$$

Since $t_*(x, y, z) = i(x, y, z-\mu)$, Prop. 14.7 holds with t_* replaced by i. Furthermore, by Lemma 14.11 we have:

$$r_{E_8} i(x, y, z) = r_{E_8} t_*(x, y, \mu + z) = m\, x \cdot y \cdot z. \tag{14.13}$$

14.14. THEOREM. *Suppose that k is perfect. For $x \in H^1(k, \mathbb{Z}/5\mathbb{Z})$ and $y, z \in H^1(k, \boldsymbol{\mu}_5)$, we have: $i(x, y, z)$ is zero in $H^1(k, E_8)$ if and only if $r_{E_8} i(x, y, z)$ is zero.*

PROOF. The "only if" direction is a basic property of the Rost invariant, so we suppose that $r_{E_8} i(x, y, z)$ is zero, i.e., that $x \cdot y \cdot z$ is zero in $H^3(k, \boldsymbol{\mu}_5^{\otimes 2})$. By the Merkurjev-Suslin Theorem, z is a reduced norm from the cyclic algebra $c_*(x, y)$, so by Example 14.6.2 we have:

$$i(x, y, z) = t_*(x, y, z + \mu) = t_*(x, y, \mu) = i(x, y, 1).$$

Now consider the class $i(x, u, 1)$ in $H^1(k(u), E_8)$ for u an indeterminate. Note that this class is split by the cyclic extension of dimension 5 defined by x and it has $r_{E_8} i(x, u, 1) = 0$ by (14.13). The proof of [**Gi 02a**, 1.4] shows that — for every completion K of $k(u)$ with respect to a discrete valuation trivial on k — the image of $i(x, u, 1)$ in $H^1(K, E_8)$ is the image of some element of $H^1(k, E_8)$, i.e., $i(x, u, 1)$ is unramified on $\mathbb{A}^1_k$. (This argument uses Bruhat-Tits theory, in particular the hypothesis that k is perfect.) We conclude that $i(x, u, 1) \in H^1(k(u), E_8)$ is also the image of a class in $H^1(k, E_8)$. By specialization, the value of $i(x, y, 1)$ does not depend on y. In particular, we have:

$$i(x, y, 1) = i(x, 1, 1) = t_*(x, 1, \mu) \quad \text{for } y \in H^1(k, \boldsymbol{\mu}_5).$$

But $c_*(x, 1)$ is the the matrix algebra $M_5(k)$, so $t_*(x, 1, \mu)$ is zero by Example 14.6.1. □

15. Mod 5 invariants of E_8

We now derive consequences of the construction in the previous section. We classify the invariants mod 5 of the split group E_8 of that type and give new examples of anisotropic groups of type E_8 over a broad class of fields. We continue the assumption that the characteristic of k_0 is $\neq 5$.

As in 14.4, the 5-torsion in $H^3(k, \mathbb{Q}/\mathbb{Z}(2))$ is identified with $H^3(k, \boldsymbol{\mu}_5^{\otimes 2})$. Composing the Rost invariant r_{E_8} with the projection on 5-torsion, we find a normalized invariant

$$h_3 \colon H^1(*, E_8) \to H^3(*, \boldsymbol{\mu}_5^{\otimes 2}).$$

15.1. THEOREM. $\mathrm{Inv}_{k_0}^{\mathrm{norm}}(E_8, \mathbb{Z}/5\mathbb{Z})$ *is a free* $R_5(k_0)$*-module with basis* h_3.

PROOF. By Cor. 3.5, we may assume that k_0 is perfect and that it contains a primitive 5-th root of unity, which we use to identify $\boldsymbol{\mu}_5$ with $\mathbb{Z}/5\mathbb{Z}$. Because i is surjective at 5, the restriction map

$$(15.2) \qquad i^* : \mathrm{Inv}^{\mathrm{norm}}(E_8, \mathbb{Z}/5\mathbb{Z}) \to \mathrm{Inv}^{\mathrm{norm}}(\mathbb{Z}/5\mathbb{Z} \times \mathbb{Z}/5\mathbb{Z} \times \mathbb{Z}/5\mathbb{Z}, \mathbb{Z}/5\mathbb{Z})$$

is an injection. By the same proof as for $\mathbb{Z}/2\mathbb{Z} \times \mathbb{Z}/2\mathbb{Z} \times \mathbb{Z}/2\mathbb{Z}$ in S16.4, we see that every normalized invariant of $\mathbb{Z}/5\mathbb{Z} \times \mathbb{Z}/5\mathbb{Z} \times \mathbb{Z}/5\mathbb{Z}$ is of the form

$$(x, y, z) \mapsto \lambda_x \cdot x + \lambda_y \cdot y + \lambda_z \cdot z + \lambda_{xy} \cdot x \cdot y + \lambda_{xz} \cdot x \cdot z + \lambda_{yz} \cdot y \cdot z + \lambda_{xyz} \cdot x \cdot y \cdot z$$

for uniquely determined λ's in $R_5(k_0)$. However, if x, y, or z is zero, then $x \cdot y \cdot z$ is zero, hence $i(x, y, z)$ is zero in $H^1(k, E_8)$ by Th. 14.14. It follows that the image of (15.2) is contained in the span of the invariant

$$(x, y, z) \mapsto \lambda_{xyz} \cdot x \cdot y \cdot z.$$

But this $R_5(k_0)$-submodule of $\mathrm{Inv}^{\mathrm{norm}}(\mathbb{Z}/5\mathbb{Z} \times \mathbb{Z}/5\mathbb{Z} \times \mathbb{Z}/5\mathbb{Z}, \mathbb{Z}/5\mathbb{Z})$ is also the submodule spanned by the restriction of h_3 by (14.13), so the theorem is proved. □

15.3. OPEN PROBLEM. (Reichstein-Youssin [**RY**, p. 1047]) Let k_0 be an algebraically closed field of characteristic zero. Do there exist nonzero invariants mapping $H^1(*, E_8)$ into $H^9(*, \mathbb{Z}/2\mathbb{Z})$ and $H^5(*, \mathbb{Z}/3\mathbb{Z})$?

15.4. COMPARISON WITH GROUPS OF TYPE F_4. There are tantalizing similarities between the behavior of groups of type F_4 relative to the prime 3 and groups of type E_8 relative to the prime 5, see e.g. [**Gi 02a**, 3.2] or compare Theorems 8.6 and 15.1. We now investigate these similarities. For E_8, the morphism i from 14.12 plays the role of the first Tits construction of Albert algebras.

As groups of type F_4 and E_8 have trivial centers and only inner automorphisms, the groups $H^1(k, F_4)$ and $H^1(k, E_8)$ are in bijection with isomorphism classes of groups of type F_4 and E_8 respectively. Using this bijection, it makes sense to write $g_3(G)$ when G is of type F_4 and g_3 denotes the invariant from 8.5, as well as to write $h_3(G)$ for G of type E_8 and h_3 as defined above.

15.5. SPLITTING BY EXTENSIONS PRIME TO p. Every group G of type F_4 over k is of the form $\mathrm{Aut}(J)$ for some Albert k-algebra J. If $g_3(G) \in H^3(k, \boldsymbol{\mu}_3^{\otimes 2})$ is nonzero, then clearly G cannot be split by an extension of dimension not divisible by 3. Conversely, if $g_3(G)$ is zero, then J is reduced, i.e., constructed from an octonion algebra O and a 2-Pfister form. In that case, every quadratic extension of k that splits O also splits J and G [**J 68**, p. 369, Th. 2].

For groups of type E_8, the analogous result is the following. It is due to Chernousov, see [**Ch 95**].

PROPOSITION. *An algebraic group G of type E_8 over k is split by an extension of k of dimension not divisible by 5 if and only if $h_3(G) = 0$.*

PROOF. As the invariant h_3 is normalized, the "only if" direction is clear. So assume that $h_3(G)$ is zero. After replacing k by an extension of dimension not divisible by 5, we may assume that k is perfect and that G equals $i(x, y, z)$ for i the map defined in 14.12 and some x, y, z. Since $r_{E_8} i(x, y, z)$ equals $h_3(G)$, Th. 14.14 gives the claim. □

15.6. ANISOTROPY. For a group $G = \mathrm{Aut}(J)$ of type F_4, one knows that G is isotropic if and only if J has nonzero nilpotents, see e.g. [**CG**, 9.1]. If $g_3(G)$ is nonzero, then J has no zero divisors (i.e., is not reduced), see [**Rost 91**] or [**PeR 96**], and in particular G is anisotropic.

We now prove the corresponding result for E_8.

PROPOSITION. *If a group G of type E_8 has $h_3(G) \neq 0$, then G is anisotropic.*

PROOF. If G is split, then clearly $h_3(G) = 0$. So suppose that G is isotropic but not split. According to the list of possible indexes in [**Tits 66**, p. 60], the semisimple anisotropic kernel A of G is a strongly inner group of type D_4, D_6, D_7, E_6, or E_7. That is, A is obtained by twisting a split simply connected group S of one of these types by a 1-cocycle $\eta \in Z^1(k, S)$. Tits's Witt-type Theorem [**Sp 98**, 16.4.2] implies that G is isomorphic to E_8 twisted by η.

The inclusion of S in E_8 comes from the obvious inclusion of Dynkin diagrams, so has Rost multiplier 1. That is, the diagram

$$\begin{array}{ccc} H^1(k, S) & \xrightarrow{r_S} & H^3(k, \mathbb{Q}/\mathbb{Z}(2)) \\ \downarrow & & \| \\ H^1(k, E_8) & \xrightarrow{r_{E_8}} & H^3(k, \mathbb{Q}/\mathbb{Z}(2)) \end{array}$$

commutes. However, for each of the possibilities for S, the Dynkin index is 2, 2, 2, 6, or 12 respectively by [**Mer**, 15.4, 16.6, 16.7], so the mod 5 portion of $r_{E_8}(\eta)$, namely $h_3(G)$, is zero. □

15.7. ANISOTROPIC GROUPS SPLIT BY EXTENSIONS OF DIMENSION p. If $G = \mathrm{Aut}(J)$ is a group of type F_4 over k where J is a first Tits construction that is not split, then G is anisotropic over k but split by a cubic extension of k. That is, nonzero symbols in $H^3(k, \boldsymbol{\mu}_3^{\otimes 2})$ give anisotropic groups of type F_4 that are split by a cubic extension.

The analogous statement for E_8 is the following:

COROLLARY (of Prop. 15.6). *If $H^3(k, \boldsymbol{\mu}_5^{\otimes 2})$ contains a nonzero symbol, then k supports an anisotropic group of type E_8 that is split by a cyclic extension of dimension* 5.

PROOF. Fix $x \in H^1(k, \mathbb{Z}/5\mathbb{Z})$ and $y, z \in H^1(k, \boldsymbol{\mu}_5)$ such that $x \cdot y \cdot z$ is not zero in $H^3(k, \boldsymbol{\mu}_5^{\otimes 2})$. Then $h_3 i(x, y, z)$ is not zero by (14.13), so the group obtained by twisting E_8 by $i(x, y, z)$ is anisotropic by Prop. 15.6. It is split by the cyclic extension of k of dimension 5 determined by x by Example 14.6. □

As a concrete example, fix a number field k. It supports a cyclic division algebra A of degree 5. (One can specify A by local data, see [**Re**, §32].) For t an indeterminate, the symbol $[A]\cdot(t)$ is nonzero in $H^3(k(t), \boldsymbol{\mu}_5^{\otimes 2})$. The proof of the corollary gives an anisotropic group of type E_8 over $k(t)$ that is split by a cyclic extension of dimension 5.

For contrast, we describe two other constructions of anisotropic groups of type E_8.

(1) Every number field k with a real embedding supports an anisotropic group of type E_8 given by Tits's construction from the octonions as in [**J 71**, p. 121]. But it follows from the Hasse Principle [**PlR**, p. 286, Th. 6.6] that an anisotropic group group of type E_8 over k cannot be split by an odd-dimensional extension of k, in contrast with the groups given by the corollary.

(2) Starting from a versal E_8-torsor, the argument in Example A.2 can be used to produce an anisotropic group of type E_8 over some field k that is split by a cyclic extension of k of dimension 5. However, this argument is not constructive and one loses choice of what k can be.

15.8. Failure of the analogy. If $G = \mathrm{Aut}(J)$ is a group of type F_4 such that the Rost invariant $r_{F_4}(G)$ is 3-torsion (i.e., $r_{F_4}(G)$ equals $g_3(G)$), then it is a result of Petersson and Racine that J is a first Tits construction [**KMRT**, 40.5]. In this case, the analogy between first Tits constructions and the map i fails. Gille [**Gi 02b**, App.] has given an example of a group G of type E_8 over a particular field k such that $r_{E_8}(G)$ is zero but G is not split. By Th. 14.14, such a G cannot be in the image of i. Similarly, the compact E_8 over the real numbers has Rost invariant zero (hence h_3 equal to zero), but clearly it is not split by an extension of degree 5, so it cannot be in the image of i.

15.9. Exercise (prime-to-5-closed fields). Suppose that k is a field such that every finite separable extension of k has dimension a power of 5. Prove that every group of type E_8 over k is split or anisotropic.

[The assumption on k is stronger than necessary; it suffices to assume that the group $H^3(k, \mathbb{Z}/6\mathbb{Z}(2))$ defined in [**Mer**, App. A] is zero.]

Part III

Spin groups

16. Introduction to Part III

In this final portion of the lectures, we determine the mod 2 invariants of the groups Spin_n, i.e., $\mathrm{Inv}_{k_0}(\mathrm{Spin}_n, \mathbb{Z}/2\mathbb{Z})$, when n is small. *Throughout Part III, we assume that k_0 — and hence every field that appears — has characteristic different from 2.* The notation Spin_n stands for the spin group of a maximally isotropic (and nondegenerate) quadratic form q_n of dimension n, see [**Kn**, §IV.6] or [**Sc**, §9.3] for details.

16.1. Spin_n FOR $n \leq 6$. In case n is very small (meaning at most 6), the group Spin_n is isomorphic to a familiar group, for which the invariants are known. We summarize these cases in Table 16. The group of normalized invariants of Spin_n is zero for $2 \leq n \leq 6$ because the set $H^1(k, \mathrm{Spin}_n)$ is zero for every k/k_0.

n	$\mathrm{Spin}_n \cong ?$	reason	$\mathrm{Inv}^{\mathrm{norm}}_{k_0}(\mathrm{Spin}_n, \mathbb{Z}/2\mathbb{Z})$
1	$\boldsymbol{\mu}_2$	from definition, see [**Kn**, p. 252]	$R_2(k_0)$, see S16.2
2	$\mathbb{G}_m$	from definition, see [**Kn**, p. 258]	0
3	SL_2	exceptional isomorphism $B_1 \cong A_1$	0
4	$\mathrm{SL}_2 \times \mathrm{SL}_2$	exceptional isomorphism $D_2 \cong A_1 \times A_1$	0
5	Sp_4	exceptional isomorphism $B_2 \cong C_2$	0
6	SL_4	exceptional isomorphism $D_3 \cong A_3$	0

TABLE 16. Description of Spin_n for $1 \leq n \leq 6$

16.2. Spin_n FOR $n \geq 7$. For n at least 7, the group Spin_n is simply connected of type B_ℓ (if n is odd, with $n = 2\ell+1$) or D_ℓ (if n is even, with $n = 2\ell$). Unfortunately, we do not have a convenient concrete description of the set $H^1(k, \mathrm{Spin}_n)$ in terms of some algebraic structure. This is in contrast to the set $H^1(k, \mathrm{SO}_n)$ for the special orthogonal group SO_n of q_n, which classifies quadratic forms of dimension n with the same determinant as q_n, see [**KMRT**, 29.29]. However, the natural action of Spin_n on the vector space underlying the quadratic form q_n gives an exact sequence

$$1 \longrightarrow \boldsymbol{\mu}_2 \longrightarrow \mathrm{Spin}_n \longrightarrow \mathrm{SO}_n \longrightarrow 1$$

which leads to an exact sequence

$$\mathrm{SO}_n(k) \xrightarrow{\delta} k^\times/k^{\times 2} \longrightarrow H^1(k, \mathrm{Spin}_n) \xrightarrow{\chi} H^1(k, \mathrm{SO}_n).$$

The image of $H^1(k, \mathrm{Spin}_n)$ in $H^1(k, \mathrm{SO}_n)$ classifies quadratic forms ϕ of dimension n such that $\phi - q_n$ belongs to I^3, see [**KMRT**, p. 437].

The map χ has zero kernel but is typically not injective. For general Galois-cohomological reasons [**Se 02**, §I.5.5], the group $k^\times/k^{\times 2}$ acts on $H^1(k, \mathrm{Spin}_n)$ such that the fibers of χ are the $k^\times/k^{\times 2}$-orbits. Further, δ is the spinor norm map [**Kn**, IV.8.3], so a twisting argument gives, for $\eta \in H^1(k, \mathrm{Spin}_n)$:

(16.3) $ck^{\times 2} \cdot \eta = \eta$ if and only if $ck^{\times 2}$ is the spinor norm of an element of $\mathrm{SO}(\chi(\eta))(k)$.

17. Surjectivities: Spin_n for $7 \leq n \leq 12$

We continue the examples of internal Chevalley modules as defined in 9.11, focusing on the case where G is Spin_n for $7 \leq n \leq 12$.

17.1. EXAMPLE ($\mathrm{Spin}_{2n-1} \cdot Z \subset \mathrm{Spin}_{2n}$). Taking $\widetilde{G}$ to be the split simply connected group of type D_{n+1} and π to be α_1, we find that G is the split simply connected group Spin_{2n} of type D_n, and V is the vector representation.

There is a G-invariant quadratic form on V and we fix an anisotropic vector v. The stabilizer of $[v]$ in $\mathrm{SO}(V)$ is easily seen to be $\boldsymbol{\mu}_2.\mathrm{SO}(v^\perp)$ (using that V is even-dimensional), hence the stabilizer of $[v]$ in G is the compositum $Z.\mathrm{Spin}_{2n-1}$, where the center Z of G meets Spin_{2n-1} in a copy of $\boldsymbol{\mu}_2$ that is the kernel of the vector representation. By dimension count, the orbit of $[v]$ is the open orbit in $\mathbb{P}(V)$.

Theorem 9.11 gives that the induced map

$$H^1(k, \mathrm{Spin}_{2n-1} \cdot Z) \to H^1(k, \mathrm{Spin}_{2n}) \tag{17.2}$$

is surjective for every k/k_0. But we can say a little more. Since Z is central, the multiplication map $\mathrm{Spin}_{2n-1} \times Z \to \mathrm{Spin}_{2n-1} \cdot Z$ is a group homomorphism, and composing this with (17.2) gives a map

$$H^1(k, \mathrm{Spin}_{2n-1}) \times H^1(k, Z) \to H^1(k, \mathrm{Spin}_{2n}) \tag{17.3}$$

and *this map is also surjective.* Indeed, the intersection $\mathrm{Spin}_{2n-1} \cap Z$ is the center of Spin_{2n-1} (a copy of $\boldsymbol{\mu}_2$), and there is an exact sequence

$$1 \longrightarrow \mathrm{Spin}_{2n-1} \longrightarrow \mathrm{Spin}_{2n-1} \cdot Z \xrightarrow{q} \boldsymbol{\mu}_2 \longrightarrow 1. \tag{17.4}$$

The center Z of Spin_{2n} satisfies

$$Z \cong \begin{cases} \boldsymbol{\mu}_4 & \text{if } n \text{ is odd,} \\ \boldsymbol{\mu}_2 \times \boldsymbol{\mu}_2 & \text{if } n \text{ is even} \end{cases}$$

and in either case the restriction of q to Z yields a surjection $H^1(k, Z) \to H^1(k, \boldsymbol{\mu}_2)$. (For surjectivity in the n odd case, see 2.5.) A twisting argument combined with the exactness of (17.4) now gives that the map

$$H^1(k, \mathrm{Spin}_{2n-1}) \times H^1(k, Z) \to H^1(k, \mathrm{Spin}_{2n-1} \cdot Z)$$

is surjective, hence that (17.3) is surjective, as claimed.

Attempting to do the same for groups of type B (equivalently, odd-dimensional quadratic forms) gives a stabilizer that is less attractive.

17.5. EXAMPLE ($G_2 \times \boldsymbol{\mu}_2 \subset \mathrm{Spin}_7$). Take $\widetilde{G}$ to be the split group of type F_4 and $\pi := \alpha_4$. The subgroup G is the split simply connected group Spin_7 of type B_3 and V is its spin representation.

Write G_2 for the split group of that type. The irreducible representation W with highest weight ω_1 is 7-dimensional (in characteristic $\neq 2$ [**GS**, p. 413]) and supports a G_2-invariant nondegenerate quadratic form q. It gives an embedding of G_2 in Spin_7. We claim that N may be taken to be the direct product of G_2 with the center $\boldsymbol{\mu}_2$ of Spin_7.

As a representation of G_2, V is a direct sum of W and a 1-dimensional representation, say $k_0 v$ [**McKP**, p. 181]. As in Example 9.12, dimension considerations imply that $[v]$ belongs to the open L-orbit in $\mathbb{P}(V)$ and G_2 is the identity component of the stabilizer N of $[v]$.

As the kernel $\boldsymbol{\mu}_2$ of the map $\mathrm{Spin}_7 \to \mathrm{SO}(W)$ clearly belongs to N, we may compute N by determining its image in $\mathrm{GL}(W)$. Since W is an irreducible representation of G_2 and every automorphism of G_2 is inner, the normalizer of G_2 in

$\mathrm{GL}(W)$ consists of scalar matrices. It follows that N is contained in $G_2.\boldsymbol{\mu}_2$, hence N equals $G_2 \times \boldsymbol{\mu}_2$.

For a version of this example over the reals, see [**Va**, Th. 3].

Combining this example with Th. 9.11 gives that every 8-dimensional form in I^3 that represents 1 is the norm quadratic form of an octonion algebra—i.e., is a 3-Pfister form—hence every 8-dimensional form in I^3 is similar to a 3-Pfister form. This is a special case of the general theorem: a 2^n-dimensional form in I^n is similar to an n-Pfister form [**Lam**, X.5.6].

17.6. EXERCISE. Prove: If q is an 8-dimensional quadratic form over k such that $C_0(q)$ is isomorphic to $M_8(K)$ for some quadratic étale k-algebra K, then q is similar to $\langle 1\rangle \oplus \langle \alpha\rangle q_0$ for $\alpha \in k^\times$ such that $K \cong k[x]/(x^2 - \alpha)$ and a uniquely determined 7-dimensional form q_0 such that $\langle 1\rangle \oplus q_0$ is a 3-Pfister form.

[This can be proved using standard quadratic form theory, or by combining Examples 17.1 and 17.5.]

In the examples above, we have used internal Chevalley modules as in 9.11 to produce representations with open orbits. For the cases where G is Spin_9 or Spin_{11}, such arguments are somewhat more complicated than the naive setup in 9.11. (See [**Ru**, 4.3(3), 5.1] for details.) Instead, we refer to Igusa's paper [**Ig**]; he proves the existence of an open orbit using concrete computations in the Clifford algebra.

17.7. EXAMPLE ($\mathrm{Spin}_7 \times \boldsymbol{\mu}_2 \subset \mathrm{Spin}_9$). As in [**Ig**, p. 1017], there are inclusions

$$\mathrm{Spin}_7 \to \mathrm{Spin}_8 \to \mathrm{Spin}_9$$

such that Spin_9 has an open orbit in $\mathbb{P}(V)$ for V its (16-dimensional) spin representation, and Spin_7 is the stabilizer of a $v \in V$ whose image in $\mathbb{P}(V)$ is in the open orbit. Recall that there are three non-conjugate embeddings of Spin_7 in Spin_8, distinguished by which copy of $\boldsymbol{\mu}_2$ in the center of Spin_8 they contain, cf. [**Dy 57a**, Th. 6.3.1] or [**Va**, Th. 5]. The $\boldsymbol{\mu}_2$ in this Spin_7 is *not* in the kernel of the map $\mathrm{Spin}_9 \to \mathrm{SO}_9$, i.e., is not the center of Spin_9.

Write Z for the copy of $\boldsymbol{\mu}_2$ that is the center of Spin_9; the element $-1 \in Z$ sends v to $-v$. But v is an anisotropic vector for the Spin_9-invariant quadratic form on V, hence $Z \times \mathrm{Spin}_7$ is the stabilizer of the line $[v]$ in Spin_9.

17.8. EXAMPLE ($G_2 \times \boldsymbol{\mu}_4 \subset \mathrm{Spin}_{10}$). Example 17.1 gives a surjection

$$H^1(k, \mathrm{Spin}_9) \times H^1(k, \boldsymbol{\mu}_4) \to H^1(k, \mathrm{Spin}_{10}).$$

Example 17.7 gives an inclusion

$$\mathrm{Spin}_7 \times \boldsymbol{\mu}_2 \subset \mathrm{Spin}_9$$

that induces a surjection on H^1's, i.e., the map

$$H^1(k, \mathrm{Spin}_7 \times \boldsymbol{\mu}_2) \times H^1(k, \boldsymbol{\mu}_4) \to H^1(k, \mathrm{Spin}_{10})$$

is surjective. The copy of $\boldsymbol{\mu}_2$ here is the center of Spin_9, which is contained in $\boldsymbol{\mu}_4$. So combining all the previous statements we obtain an inclusion

$$\mathrm{Spin}_7 \times \boldsymbol{\mu}_4 \subset \mathrm{Spin}_{10}$$

that gives a surjection on H^1's.

In terms of quadratic forms, we view Spin_{10} as the spin group of the quadratic form $q := \langle 1, -1\rangle \oplus 4\langle 1, -1\rangle$, where Spin_7 acts on the second summand. Therefore, we have proved that the image of the map

$$H^1(k, \mathrm{Spin}(q)) \to H^1(k, \mathrm{SO}(q)) \tag{17.9}$$

consists of isotropic quadratic forms. On the other hand, the image of (17.9) is precisely the collection of 10-dimensional forms in I^3, so we have recovered Pfister's result—see [**Pf**, p. 123] or [**Lam**, XII.2.8]—that *such forms are isotropic.* (Pfister's proof used quadratic form theory. Tits gave a characteristic-free proof using algebraic groups in [**Tits 90**, 4.4.1(ii)]. We remark that this theorem has been generalized by Hoffmann, Vishik, and Karpenko: There are no anisotropic forms in I^n of dimension d such that $2^n < d < 2^n + 2^{n-1}$ for $n \geq 2$, see e.g. [**EKM**].)

We can find a subgroup of Spin_{10} that is smaller than $\mathrm{Spin}_7 \times \boldsymbol{\mu}_4$ and yet still gives a surjection on H^1's. As in the remarks at the end of Example 17.5, everything in the image of (17.9) is in the image of

$$H^1(k, G_2 \times \boldsymbol{\mu}_4) \to H^1(k, \mathrm{Spin}(q)) \to H^1(k, \mathrm{SO}(q)), \tag{17.10}$$

where G_2 is a subgroup of Spin_7 in the natural way. Said differently, everything in $H^1(k, \mathrm{Spin}(q))$ is in the $H^1(k, \boldsymbol{\mu}_4)$-orbit of something in the image of $H^1(k, G_2)$, i.e., the first map in (17.10) is surjective.

Instead of starting with Example 17.1, we could have viewed Spin_{10} as a subgroup of E_6, in which case the representation V given by 9.11 is a half-spin representation. However, this gives an ugly stabilizer, see [**Ig**, Prop. 2].

The following exercise gives an example of a useful surjection on cohomology.

17.11. Exercise. Recall that for every quadratic étale k_0-algebra k_1, there is a surjective functor $\mathrm{Quad}_n \to \mathrm{Herm}_{k_1/k_0,n}$ that sends a quadratic form q to a k_1/k_0-hermitian form q_H ("hermitian forms can be diagonalized"). That is, in the commutative diagram

$$\begin{array}{ccc} H^1(*, \mathrm{O}(q)) & \longrightarrow & H^1(*, \mathrm{U}(q_H)) \\ \uparrow & & \uparrow \\ H^1(*, \mathrm{SO}(q)) & \longrightarrow & H^1(*, \mathrm{SU}(q_H)), \end{array}$$

the top arrow is a surjection. Prove that the bottom arrow is also a surjection.

17.12. Example ($\mathrm{SO}_6 \times \boldsymbol{\mu}_4 \subset \mathrm{Spin}_{12}$). Take $\widetilde{G}$ to be the split simply connected group of type E_7 and $\pi := \alpha_1$. The subgroup G is the split simply connected group Spin_{12} of type D_6 and V is a half-spin representation. Speaking concretely, we view Spin_{12} as the spin group of the quadratic form $x \mapsto b(x,x)$ on k_0^{12} where

$$b(x,y) := \frac{1}{2} x^t \left(\begin{smallmatrix} 0 & \delta_6 \\ \delta_6 & 0 \end{smallmatrix}\right) y \qquad (x, y \in k_0^{12})$$

and δ_n denotes the diagonal n-by-n matrix with (j,j)-entry $(-1)^j$. The element $v := 1 + e_1 e_2 \cdots e_6 \in V$—where e_j denotes the basis vector of k_0^{12} whose only nonzero entry is a 1 in the j-th place—belongs to the open orbit in $\mathbb{P}(V)$, cf. [**Ig**]. The stabilizer of the vector v is isomorphic to SL_6 such that the vector representation $\chi \colon \mathrm{Spin}_{12} \to \mathrm{SO}_{12}$ restricts to the map

$$a \mapsto \left(\begin{smallmatrix} a & 0 \\ 0 & \delta_6 a^{-t} \delta_6 \end{smallmatrix}\right) \qquad (a \in \mathrm{SL}_6),$$

cf. [**Ig**, Prop. 3].

We compute the stabilizer N of the line $[v]$. Fix a primitive 4-th root of unity i (in some algebraic closure of k_0) and put

$$s := i\,(e_1 + e_7)(e_2 + e_8)(e_3 + e_9)(e_4 + e_{10})(e_5 + e_{11})(e_6 + e_{12}).$$

Because

$$(e_j + e_{6+j})^2 = 2b(e_j, e_{6+j}) = (-1)^j \quad (1 \leq j \leq 6),$$

we have

$$s^2 = i^2 \cdot (-1)^3 \cdot (-1)^{\binom{6}{2}} = -1,$$

the nonidentity element in the kernel of the vector representation of Spin_{12}. Hence s^{-1} is $-s$, equivalently, s with the order of terms in the product reversed. For $1 \le j \le 6$, we have:

$$(e_j + e_{6+j})e_\ell(e_j + e_{6+j}) = \begin{cases} (-1)^j e_{6+j} & \text{if } \ell = j \\ (-1)^j e_j & \text{if } \ell = 6+j \\ -(-1)^j e_\ell & \text{otherwise} \end{cases}$$

It is now easy to see that conjugation by s acts on k_0^{12} by the matrix

$$\chi(s) = \left(\begin{smallmatrix} 0 & -1 \\ -1 & 0 \end{smallmatrix}\right).$$

This matrix is in SO_{12}, so s belongs to Spin_{12}.

Finally, one checks that $sv = i\,v$. Since V supports a Spin_{12}-invariant quartic form [**Ig**, Prop. 3], it follows that N is generated by SL_6 and s and in particular is isomorphic to $\mathrm{SL}_6 \rtimes \boldsymbol{\mu}_4$ by identifying s with i. The equation

$$\chi(sas^{-1}) = \chi(s)\chi(a)\chi(s)^{-1} = \chi(\delta_6 a^{-t}\delta_6)$$

implies that $sas^{-1} = \delta_6 a^{-t} \delta_6$ for $a \in \mathrm{SL}_6$.

As in Th. 9.11 (cf. Example 12.13) we have a surjection

$$H^1(*, \mathrm{SL}_6 \rtimes \boldsymbol{\mu}_4) \to H^1(*, \mathrm{Spin}_{12}).$$

Write SO_6 for the special orthogonal group of the hyperbolic symmetric bilinear form $(x, y) \mapsto x^t \delta_6 y$. It is a subgroup of SL_6 and is fixed elementwise by the map $a \mapsto \delta_6 a^{-t} \delta_6$, so there is a natural inclusion

$$\mathrm{SO}_6 \times \boldsymbol{\mu}_4 \hookrightarrow \mathrm{SL}_6 \rtimes \boldsymbol{\mu}_4$$

that is the identity on $\boldsymbol{\mu}_4$. The induced map

$$H^1(k, \mathrm{SO}_6 \times \boldsymbol{\mu}_4) \to H^1(k, \mathrm{SL}_6 \rtimes \boldsymbol{\mu}_4)$$

is a surjection for every extension k/k_0. (To see that a given class $\eta \in H^1(k, \mathrm{SL}_6 \rtimes \boldsymbol{\mu}_4)$ is in the image, twist by the image of η in $H^1(k, \boldsymbol{\mu}_4)$ and then apply Exercise 17.11.) It follows that *the inclusion* $\mathrm{SO}_6 \times \boldsymbol{\mu}_4 \to \mathrm{Spin}_{12}$ *induces a surjection on* H^1*'s.*

Concretely, the preceding example shows:

17.13. THEOREM (Pfister [**Pf**, pp. 123, 124]). *Every 12-dimensional quadratic form in* I^3 *can be written as* $\langle d \rangle \langle 1, -c \rangle (\phi_1' \oplus \langle -1 \rangle \phi_2')$ *for some* $c, d \in k^\times$ *and 2-Pfister quadratic forms* ϕ_1, ϕ_2.

Here we have written ϕ_j' for the pure part of ϕ_j, i.e., for the unique form such that $\phi_j = \langle 1 \rangle \oplus \phi_j'$.

PROOF. A 12-dimensional quadratic form q in I^3 over k is represented by a class $\eta \in H^1(k, \mathrm{SO}_{12})$ that is in the image of $H^1(k, \mathrm{Spin}_{12})$. Example 17.12 shows that η is equivalent to a class coming from $H^1(k, \mathrm{SO}_6 \times \boldsymbol{\mu}_4)$. That is, q is isomorphic to $\langle 1, -c \rangle q_6$ for some 6-dimensional form q_6 of determinant -1. But such a ψ is of the form $\langle d \rangle (\phi_1' \oplus \langle -1 \rangle \phi_2')$, i.e., is an Albert form, see e.g. [**KMRT**, 16.3] or [**Lam**, XII.2.13]. □

Hoffmann has conjectured [**Ho**, Conj. 2] a generalization of Th. 17.13 for forms of dimension $2^n + 2^{n-1}$ in I^n with $n \ge 4$.

17.14. EXAMPLE ($\mathrm{SO}_5 \times \boldsymbol{\mu}_4 \subset \mathrm{Spin}_{11}$). We view Spin_{12} as the spin group of the bilinear form b from Example 17.12. In this way, we see Spin_{11} as a subgroup of Spin_{12} consisting of elements that fix the vector

$$\varepsilon_1 := e_6 - e_{12}$$

in the space underlying b. The image of Spin_{11} under the vector representation $\chi\colon \mathrm{Spin}_{12} \to \mathrm{SO}(b)$ is the special orthogonal group of b restricted to the subspace with basis

$$\varepsilon_0 := e_6 + e_{12}, e_1, e_2, \ldots, e_5, e_7, e_8, e_9, \ldots, e_{11}.$$

The spin representation of Spin_{11} is the restriction of the half-spin representation V of Spin_{12} from the previous example, and $v := 1 + e_1e_2\cdots e_6$ is again a representative of the open orbit in $\mathbb{P}(V)$ [**Ig**, Prop. 6]. The stabilizer N of $[v]$ in Spin_{11} is the intersection of Spin_{11} with $\mathrm{SL}_6 \rtimes \boldsymbol{\mu}_4$.

The intersection of $\chi(\mathrm{Spin}_{11})$ with $\chi(\mathrm{SL}_6 \rtimes \boldsymbol{\mu}_4)$ in $\mathrm{SO}(b)$ is the subgroup of $\chi(\mathrm{SL}_6 \rtimes \boldsymbol{\mu}_4)$ that fixes ε_1. Write I_n for the n-by-n identity matrix. The intersection is $\mathrm{SL}_5 \rtimes \boldsymbol{\mu}_2$, where SL_5 is viewed as the subgroup of matrices

$$\left(\begin{array}{cc|cc} a & & & \\ & I_1 & & \\ \hline & & \delta_5 a^{-t}\delta_5 & \\ & & & I_1 \end{array}\right) \qquad (a \in \mathrm{SL}_5)$$

of $\mathrm{SO}(b)$ and the nontrivial element of $\boldsymbol{\mu}_2$ is $\chi(s)$ for s from Example 17.12. It follows that the stabilizer N of $[v]$ in Spin_{11} is $\mathrm{SL}_5 \rtimes \boldsymbol{\mu}_4$. As in the previous example, the composition

$$\mathrm{SO}_5 \times \boldsymbol{\mu}_4 \to \mathrm{SL}_5 \rtimes \boldsymbol{\mu}_4 \to \mathrm{Spin}_{11}$$

induces a surjection on H^1's.

18. Invariants of Spin_n for $7 \le n \le 10$

We now determine the invariants of Spin_n (for $7 \le n \le 10$) with values in $\mathbb{Z}/2\mathbb{Z}$. These results have also been obtained by Mark MacDonald by somewhat different means, see [**MacD**].

18.1. INVARIANTS OF Spin_8. Combining Examples 17.5 and 17.1, we find an inclusion

$$i\colon G_2 \times Z \to \mathrm{Spin}_8$$

such that the induced map i_* on H^1's is surjective, where Z is the center of Spin_8 and is isomorphic to $\boldsymbol{\mu}_2 \times \boldsymbol{\mu}_2$. Fix inequivalent 8-dimensional irreducible representations $\chi_1, \chi_2\colon \mathrm{Spin}_8 \to \mathrm{SO}_8$ such that χ_j restricts to be the projection $Z \to \boldsymbol{\mu}_2$ on the j-th factor; they induce invariants $\underline{\chi}_j\colon H^1(*, Z) \to H^1(*, \boldsymbol{\mu}_2)$.

As Spin_8 is split, the image of $H^1(k, Z)$ in $H^1(k, \mathrm{Spin}_8)$ is zero. Applying Lemma 6.7, i^* identifies $\mathrm{Inv}^{\mathrm{norm}}_{k_0}(\mathrm{Spin}_8, \mathbb{Z}/2\mathbb{Z})$ with an $R_2(k_0)$-submodule of the free module I with basis the invariants

$$e_3, \quad e_3\cdot\underline{\chi}_1, \quad e_3\cdot\underline{\chi}_2, \quad e_3\cdot\underline{\chi}_1\cdot\underline{\chi}_2 \quad \in \mathrm{Inv}^{\mathrm{norm}}_{k_0}(G_2 \times Z, \mathbb{Z}/2\mathbb{Z}). \tag{18.2}$$

We prove that each of the invariants in (18.2) is the restriction of an invariant of Spin_8.

Let (C, ζ) be a class in $H^1(k, G_2 \times Z)$, where C is an octonion algebra. Abusing notation, we write $\chi_j(\zeta)$ for the corresponding element of $k^\times/k^{\times 2}$. The composition

$$H^1(k, G_2 \times Z) \longrightarrow H^1(k, \mathrm{Spin}_8) \xrightarrow{\chi_j} H^1(k, \mathrm{SO}_8) \tag{18.3}$$

sends (C,ζ) to the quadratic form $\langle\chi_j(\zeta)\rangle N_C$, where N_C denotes the norm on C. Composing (18.3) with the Arason invariant e_3 defined in Example 1.2.3 sends (C,ζ) to $e_3(C)$. That is, the invariant e_3 from (18.2) is the restriction of an invariant of Spin_8.

Of course, $e_3(C)$ is zero whenever $\langle\chi_j(\zeta)\rangle N_C$ is isotropic. Applying Prop. 10.2 to the representations χ_1 and χ_2, we find that $e_3\cdot\underline{\chi}_1$ and $e_3\cdot\underline{\chi}_2$ are also restrictions of invariants of Spin_8. Finally, $e_3\cdot\underline{\chi}_1$ is zero whenever $\langle\chi_2(\zeta)\rangle N_C$ is isotropic, and applying Prop. 10.2 again gives that $e_3\cdot\underline{\chi}_1\cdot\underline{\chi}_2$ is the restriction of an invariant of Spin_8.

We have proved that $\mathrm{Inv}^{\mathrm{norm}}_{k_0}(\mathrm{Spin}_8,\mathbb{Z}/2\mathbb{Z})$ is a free $R_2(k_0)$-module of rank 4 with generators of degree 3, 4, 4, 5.

18.4. Exercise. Show: For every Pfister form ϕ and every $c\in k^\times$, the group of spinor norms of elements of $\mathrm{SO}(\langle c\rangle\phi)(k)$ is the same as the group of $\lambda\in k^\times$ such that $\langle\lambda\rangle\phi$ is isomorphic to ϕ.

[It suffices for ϕ to be round in the sense of [**Lam**, V.1.13].]

18.5. Remark. The invariants of Spin_8 distinguish elements of $H^1(k,\mathrm{Spin}_8)$. More precisely, writing a_j for the invariant of Spin_8 that restricts to $e_3\cdot\underline{\chi}_j$ on Z (in the notation of 18.1), we claim that the function

$$H^1(k,\mathrm{Spin}_8)\xrightarrow{e_3\times a_1\times a_2}H^3(k,\mathbb{Z}/2\mathbb{Z})\times H^4(k,\mathbb{Z}/2\mathbb{Z})\times H^4(k,\mathbb{Z}/2\mathbb{Z})$$

is injective.

Suppose that $e_3\times a_1\times a_2$ takes the same value on classes $(C,\alpha,\beta),(C',\alpha',\beta')$ in $H^1(k,G_2\times Z)$. As $e_3(C,\alpha,\beta)$ equals $e_3(C',\alpha',\beta')$, clearly C equals C'. Further,

$$0=a_1(C,\alpha,\beta)-a_1(C,\alpha',\beta')=(\alpha\alpha')\cdot e_3(C),$$

so $\alpha\alpha'$ is a similarity factor of C and a spinor norm for $\langle\beta'\rangle N_C$ by Exercise 18.4. Applying (16.3), we see that (C,α',β') and (C,α,β') have the same image in $H^1(k,\mathrm{Spin}_8)$. Repeating this argument with obvious changes gives that (C,α',β') and (C,α,β) have the same image in $H^1(k,\mathrm{Spin}_8)$, proving the claim.

18.6. Invariants of Spin_7. By Example 17.5, there is a subgroup $G_2\times\boldsymbol{\mu}_2$ of Spin_7 such that the induced map

$$i_*\colon H^1(*,G_2\times\boldsymbol{\mu}_2)\to H^1(*,\mathrm{Spin}_8)$$

is surjective. Combined with the inclusion $\mathrm{Spin}_7\hookrightarrow\mathrm{Spin}_8$ obtained by viewing Spin_7 as the identity component of the stabilizer of a vector of length 1, we have maps

$$\mathrm{Inv}^{\mathrm{norm}}_{k_0}(G_2\times\boldsymbol{\mu}_2,\mathbb{Z}/2\mathbb{Z})\hookleftarrow\mathrm{Inv}^{\mathrm{norm}}_{k_0}(\mathrm{Spin}_7,\mathbb{Z}/2\mathbb{Z})\leftarrow\mathrm{Inv}^{\mathrm{norm}}_{k_0}(\mathrm{Spin}_8,\mathbb{Z}/2\mathbb{Z}).$$

As in 18.1, the image in $\mathrm{Inv}^{\mathrm{norm}}_{k_0}(G_2\times\boldsymbol{\mu}_2,\mathbb{Z}/2\mathbb{Z})$ is contained in the free $R_2(k_0)$-module I with basis $e_3, e_3\cdot\underline{\mathrm{id}}$ and the invariant e_3 is the restriction of the Arason invariant on SO_8. Similarly, the copy of $\boldsymbol{\mu}_2$ in Spin_7 is the kernel of a representation $\mathrm{Spin}_8\to\mathrm{SO}_8$, say χ_2. The invariant $e_3\cdot\underline{\chi}_1$ of Spin_8 restricts to the invariant $e_3\cdot\underline{\mathrm{id}}$ of $G_2\times\boldsymbol{\mu}_2$. This proves that $\mathrm{Inv}^{\mathrm{norm}}_{k_0}(\mathrm{Spin}_7,\mathbb{Z}/2\mathbb{Z})$ is a free $R_2(k_0)$-module of rank 2 with basis elements of degrees 3 and 4.

18.7. Invariants of Spin_{10}. From Example 17.8 we have an inclusion $i:G_2\times\boldsymbol{\mu}_4\to\mathrm{Spin}_{10}$ such that the induced map i_* on H^1's is surjective. For a Cayley k-algebra C and $\alpha\in k^\times/k^{\times 4}$, define

$$a_3(C,\alpha)=e_3(C)\quad\text{and}\quad a_4(C,\alpha)=e_3(C)\cdot\underline{s}(\alpha)$$

in $\mathrm{Inv}^{\mathrm{norm}}_{k_0}(G_2 \times \boldsymbol{\mu}_4, \mathbb{Z}/2\mathbb{Z})$. (The invariant $\underline{s}$ is defined in 2.5.) As for Spin_8, i^* identifies $\mathrm{Inv}^{\mathrm{norm}}_{k_0}(\mathrm{Spin}_{10}, \mathbb{Z}/2\mathbb{Z})$ with a submodule of the free $R_2(k_0)$ module with basis a_3, a_4. The image of a pair (C, α) in $H^1(k, \mathrm{SO}_{10})$ corresponds to the quadratic form $\langle 1, -1\rangle \oplus \langle \alpha \rangle N_C$, so a_3 and a_4 are obviously restrictions of invariants of Spin_{10}.

18.8. EXERCISE. Prove:

(1) The function $H^1(k, \mathrm{Spin}_{10}) \to H^1(k, \mathrm{SO}_{10})$ induced by the map $\mathrm{Spin}_{10} \to \mathrm{SO}_{10}$ is injective.

[Of course, this statement is much stronger than "has zero kernel".]

(2) The invariants of Spin_{10} distinguish elements of $H^1(k, \mathrm{Spin}_{10})$, i.e., the function

$$H^1(k, \mathrm{Spin}_{10}) \xrightarrow{a_3 \times a_4} H^3(k, \mathbb{Z}/2\mathbb{Z}) \times H^4(k, \mathbb{Z}/2\mathbb{Z})$$

is injective.

18.9. INVARIANTS OF Spin_9. We view $\mathrm{Spin}_8 \subset \mathrm{Spin}_9 \subset \mathrm{Spin}_{10}$ as the spin groups of the quadratic forms

$$4\langle 1, -1\rangle, \quad \langle -1 \rangle \oplus 4\langle 1, -1\rangle, \quad \text{and} \quad \langle 1, -1\rangle \oplus 4\langle 1, -1\rangle$$

in the obvious manner. Combining Examples 17.5 and 17.7 and putting $Z = \boldsymbol{\mu}_2 \times \boldsymbol{\mu}_2$, we find an inclusion of $G_2 \times Z$ in Spin_9 that gives a surjection

$$H^1(*, G_2 \times Z) \to H^1(*, \mathrm{Spin}_9)$$

and identifies $\mathrm{Inv}^{\mathrm{norm}}_{k_0}(\mathrm{Spin}_9, \mathbb{Z}/2\mathbb{Z})$ with a submodule of $\mathrm{Inv}^{\mathrm{norm}}_{k_0}(G_2 \times Z, \mathbb{Z}/2\mathbb{Z})$, contained in the free $R_2(k_0)$-module with basis (18.2).

For the sake of fixing notation, suppose that the restriction of the vector representation of Spin_9 to Spin_8 is the direct sum of χ_1 (as opposed to χ_2 or χ_3) and a 1-dimensional trivial representation. The image of a pair $(C, \zeta) \in H^1(k, G_2 \times Z)$ under the maps

$$H^1(k, G_2 \times Z) \to H^1(k, \mathrm{Spin}_8) \to H^1(k, \mathrm{Spin}_{10}) \to H^1(k, \mathrm{SO}_{10})$$

is $\langle 1, -1\rangle \oplus \langle \chi_1(\zeta)\rangle N_C$. Thus the invariants a_3 and a_4 of Example 18.7 restrict to invariants e_3 and $e_3 \cdot \underline{\chi}_1$ of $G_2 \times Z$ from (18.2).

We can also view Spin_9 as the subgroup of the automorphism group of the split Albert algebra J consisting of the algebra automorphisms that fix a primitive idempotent in J [**J 68**, §IX.3]. Restricting J to a representation of Spin_8, we find a direct sum of a 3-dimensional trivial representation and the three inequivalent irreps $\chi_1, \chi_2, \chi_3 \colon \mathrm{Spin}_8 \to \mathrm{SO}_8$. The invariant

$$f_5 \colon H^1(*, \mathrm{Aut}(J)) \to H^5(*, \mathbb{Z}/2\mathbb{Z})$$

defined in §22 of S restricts to be nonzero on $G_2 \times Z$. If -1 is a square in k_0, then its restriction is the invariant $e_3 \cdot \underline{\chi}_1 \cdot \underline{\chi}_2$ on $G_2 \times Z$ from (18.2). (The assumption on -1 is here only for the convenience of ignoring various factors of -1.)

Finally we claim that the invariant $\lambda \cdot e_3 \cdot \underline{\chi}_2$ of $G_2 \times Z$, for every nonzero $\lambda \in R_2(k_0)$, is *not* the restriction of an invariant of Spin_9. Let k be the extension of k_0 obtained by adjoining indeterminates x, y, z, w, and write C for the Cayley k-algebra with $e_3(C)$ equal to $(x) \cdot (y) \cdot (z)$. Fix a $\zeta \in H^1(k, Z)$ such that $\chi_1(\zeta) = (1)$ and $\chi_2(\zeta) = (w)$. The invariant $e_3 \cdot \underline{\chi}_2$ takes different values on $(C, 1)$ and (C, ζ), namely 0 and $\lambda \cdot (x) \cdot (y) \cdot (z) \cdot (w)$. However, the two classes have the same image in $H^1(k, \mathrm{SO}_9)$, the form $\langle -1 \rangle \oplus N_C$. As this form is isotropic, its spinor norm map is onto and the fiber of

$$H^1(k, \mathrm{Spin}_9) \to H^1(k, \mathrm{SO}_9)$$

over $\langle -1\rangle \oplus N_C$ is a singleton by (16.3). That is, $(C,1)$ and (C,w) have the same image in $H^1(k,\mathrm{Spin}_9)$, proving the claim.

In the case where -1 is a square in k_0, this determines $\mathrm{Inv}^{\mathrm{norm}}_{k_0}(\mathrm{Spin}_9,\mathbb{Z}/2\mathbb{Z})$: it is free of rank 3 with basis elements of degree 3, 4, 5.

19. Divided squares in the Grothendieck-Witt ring

In this section, we define a function $P_n\colon I^n \to I^{2n}$ in the Witt ring that will be used to construct invariants of Spin_n for $n = 11, 12, 14$. It can also be used to give bounds on the symbol length of a class in $H^d(k,\mathbb{Z}/2\mathbb{Z})$, cf. Example A.3.

Recall the Grothendieck-Witt ring $\widehat{W}$ (denoted WGr in S) over a field k of characteristic $\neq 2$: it is the ring of formal differences of (nondegenerate) quadratic forms over k. It is a λ-ring in the sense of Grothendieck, see e.g. S27.1. For a quadratic form $q = \langle \alpha_1, \alpha_2, \ldots, \alpha_n\rangle$ and $0 < p \leq n$, we have

$$\lambda^p q = \oplus_{i_1<i_2<\cdots<i_p} \langle \alpha_{i_1}\alpha_{i_2}\cdots\alpha_{i_p}\rangle.$$

In particular, $\lambda^0 q = \langle 1\rangle$ and $\lambda^1 q = q$.

19.1. Example. Writing $\mathcal{H}$ for a hyperbolic plane $\langle 1,-1\rangle$, we have:

$$\lambda^2(n\mathcal{H}) \cong (n^2-n)\mathcal{H} \oplus n\langle -1\rangle.$$

19.2. Exercise. Prove: The Killing form on the Lie algebra $\mathfrak{so}(q)$ is isomorphic to $\langle -2\rangle\langle \dim q - 2\rangle\lambda^2 q$.

We will only make use of λ^2. Here are a few useful identities in $\widehat{W}$, where x and y denote quadratic forms:

$$\lambda^2(x+y) = \lambda^2 x + xy + \lambda^2 y \tag{19.3}$$

$$\lambda^2(\langle c\rangle x) = \lambda^2 x \tag{19.4}$$

$$\lambda^2(x-y) = \lambda^2 x - y(x-y) - \lambda^2 y = \lambda^2 x - xy + \dim y + \lambda^2 y \tag{19.5}$$

$$\lambda^2(xy) = x^2\lambda^2 y + y^2\lambda^2 x - 2(\lambda^2 x)(\lambda^2 y) \tag{19.6}$$

19.7. Example. For a quadratic form z and a natural number n, the image of $\lambda^2(z-n\mathcal{H})$ in the Witt ring is $n + \lambda^2 z$, as can be seen by combining (19.5) and Example 19.1.

19.8. Lemma. *For every n-Pfister form ϕ with $n \geq 1$, we have: $\lambda^2\phi \cong 2^{n-1}\phi'$.*

Proof. By induction on n. As $\lambda^2\langle 1,-\alpha\rangle$ is isomorphic to $\langle -\alpha\rangle$, the case $n=1$ holds. For ϕ an n-Pfister form with $n>1$, we may write[j] $\phi = \langle\langle \alpha\rangle\rangle\psi$ for some $\alpha \in k^\times$ and $(n-1)$-Pfister ψ. In $\widehat{W}$, we have

$$\lambda^2\phi = \langle\langle \alpha\rangle\rangle^2\lambda^2\psi + \langle -\alpha\rangle\psi^2 - 2\langle -\alpha\rangle\lambda^2\psi$$

by (19.6). In the Witt ring, $\langle\langle \alpha\rangle\rangle^2 - 2\langle -\alpha\rangle$ equals 2, and

$$\lambda^2\phi = 2\lambda^2\psi + \langle -\alpha\rangle\psi^2,$$

which by the induction hypothesis is

$$2^{n-1}\psi' + \langle -\alpha\rangle\psi^2 = 2^{n-1}(\psi' + \langle -\alpha\rangle\psi) = 2^{n-1}(\langle\langle \alpha\rangle\rangle\psi)'.$$

Since $\lambda^2\phi$ equals $2^{n-1}\phi'$ in the Witt ring and both have dimension $2^{n-1}(2^n-1)$, the conclusion follows. □

---

[j]Here and below we write $\langle\langle \alpha_1,\ldots,\alpha_n\rangle\rangle$ for the n-Pfister form $\otimes_{i=1}^n\langle 1,-\alpha_i\rangle$.

For q an even-dimensional quadratic form, there is a canonical lift $\hat{q}$ to the Grothendieck-Witt ring $\widehat{W}$, namely

$$\hat{q} := q - r\mathcal{H}, \quad \text{where } \dim q = 2r. \tag{19.9}$$

Note that $\hat{q} \in \widehat{W}$ only depends on q up to Witt-equivalence. (This is just a restatement of the fact that the quotient map $\widehat{W} \to W$ restricts to an isomorphism $\widehat{I} \xrightarrow{\sim} I$, where $\widehat{I}$ is ideal of zero-dimensional virtual forms; $\hat{q}$ is the inverse image of q under this isomorphism.) For $n \geq 1$, we define

$$P_n \colon I \to W \quad \text{via} \quad P_n(x) := \lambda^2 \hat{x} - 2^{n-1}x,$$

where we conflate $\lambda^2\hat{x}$ with its image in the Witt ring. We remark that the device of replacing x with $\hat{x}$ is necessary, as λ^2 is not well-behaved with respect to Witt-equivalence. (For example, the dimensions of $\lambda^2\mathcal{H}$ and $\lambda^2(2\mathcal{H})$ are not even congruent mod 2.)

Using Example 19.7, it is easy to check that

$$P_n(x+y) = P_n(x) + xy + P_n(y) \tag{19.10}$$

and

$$P_n(\langle c\rangle x) = P_n(x) + 2^{n-1}\langle\langle c\rangle\rangle x \tag{19.11}$$

hold, for $x, y \in I$ and $c \in k^\times$.

19.12. PROPOSITION. *For $n \geq 1$:*

(1) *P_n is zero on n-Pfister forms.*
(2) *P_n restricts to a map $I^n \to I^{2n}$.*

If $n \geq 2$ and -1 is a square in k:

(3) *P_n induces a map $I^n/I^{n+1} \to I^{2n}/I^{2n+1}$.*
(4) *For $c_i \in k^\times$ and n-Pfister forms ϕ_i, we have:*

$$P_n\left(\sum_i \langle c_i\rangle\phi_i\right) = \sum_{i<j} \langle c_i c_j\rangle \phi_i\phi_j$$

PROOF. Combining Example 19.7 and Lemma 19.8 gives (1). For (2), we use that every element of I^n is a sum of elements of the form $\langle c\rangle\phi$, where ϕ is an n-Pfister form and c is in $k^\times$. By (19.10) and (19.11), it suffices to prove that $P_n(\phi)$ belongs to I^{2n}, which is true by (1).

Both (3) and (4) rest on the fact that $2^{n-1} = 0$ in the Witt ring because n is at least 2 and -1 is a square. We prove (3). Let $x, y \in I^n$ be such that $z := x - y$ belongs to I^{n+1}. Then $\hat{z} = \hat{x} - \hat{y}$ in $\widehat{W}$, and we have

$$P_{n+1}(z) = \lambda^2\hat{x} - \lambda^2\hat{y} - yz - 2^n z$$

by (19.5). So:

$$P_n(x) - P_n(y) = P_{n+1}(z) + yz + 2^{n-1}z.$$

All three summands on the right belong to I^{2n+1}. For the first term, this is (2). For the last term, it is because $2^{n-1} = 0$.

As for (4), under our special hypotheses, Equation (19.11) takes the nice form:

$$P_n(\langle c\rangle x) = P_n(x).$$

Applying (19.10) and (1) gives (4). □

For the remainder of this section, we maintain the hypotheses that -1 is a square in k_0 and n is at least 2. Applying the map e_{2n} from Example 1.2.3 to Prop. 19.12.4 gives:

$$e_{2n}\left(P_n\left(\sum\nolimits_i \langle c_i\rangle\phi_i\right)\right) = \sum_{i<j} e_n(\phi_i)e_n(\phi_j). \tag{19.13}$$

19.14. EXAMPLE (Invariants of SO(6)). We write the invariants of SO(6) (the special orthogonal group of the dot product on k_0^6) in terms of the maps e_n and P_n. By S20.6, the normalized invariants of SO(6) with values in $\mathbb{Z}/2\mathbb{Z}$ form a free $R_2(k_0)$-module with basis w_2, w_4, b, where b satisfies

$$b(\langle\alpha_1, \alpha_2, \ldots, \alpha_5, \alpha_6\rangle) = (\alpha_1)\cdot(\alpha_2)\cdots(\alpha_5).$$

(In S20.1, this b was denoted "b_1", where 1 is the nonzero element of $H^0(k_0, \mathbb{Z}/2\mathbb{Z})$, i.e., the identity element of $R_2(k_0)$.)

An element of $H^1(k, \mathrm{SO}(6))$ corresponds to a 6-dimensional form q in I^2. Such a form is isomorphic to $\langle\beta\rangle(\phi_1' \oplus \langle -1\rangle\phi_2')$ for some $\beta \in k^\times$ and 2-Pfister forms ϕ_1, ϕ_2. Direct computation gives

$$w_2(q) = e_2(q),$$
$$w_4(q) = w_2(\langle\beta\rangle\phi_1')\cdot w_2(\langle\beta\rangle\phi_2') = e_2(\phi_1)\cdot e_2(\phi_2) = e_4(P_2(q))$$

by (19.13), and

$$b(q) = (\beta)\cdot w_4(q).$$

19.15. Prop. 19.12.4 makes P_n look like a "divided square", meaning a squaring operation from a divided power structure. We remark that—still assuming that -1 is a square—there are also divided square operations on Milnor K-theory $P_n^M : K_n^M/2 \to K_{2n}^M/2$, see [**Ka**, App. A]. For $n \geq 2$, the diagram

$$\begin{array}{ccc} K_n^M/2 & \xrightarrow{P_n^M} & K_{2n}^M/2 \\ \downarrow & & \downarrow \\ I^n/I^{n+1} & \xrightarrow{P_n} & I^{2n}/I^{2n+1} \end{array}$$

commutes, where the vertical arrows are the natural surjections that send the symbol $\{\alpha_1, \alpha_2, \ldots, \alpha_n\}$ to the class of the Pfister form $\langle\langle\alpha_1, \alpha_2, \ldots, \alpha_n\rangle\rangle$.

20. Invariants of Spin_{11} and Spin_{12}

We now determine the invariants of Spin_{12} and Spin_{11} with values in $\mathbb{Z}/2\mathbb{Z}$. We begin with some results on quadratic forms.

20.1. LEMMA. *Let x, y be quadratic forms of the same dimension and fix $c \in k^\times$. if $\langle\langle c\rangle\rangle(x - y)$ is zero in the Witt ring, then $\langle\langle c\rangle\rangle\lambda^2(x - y) \in \widehat{W}$ maps to zero in the Witt ring.*

PROOF. Replacing x, y with $x \oplus \langle -1\rangle y$, $(\dim y)\mathcal{H}$ respectively does not change $\langle\langle c\rangle\rangle(x - y)$ nor the image of $\lambda^2(x - y)$ in the Witt ring. Therefore, we may assume that x has even dimension $2r$ and $y = r\mathcal{H}$.

The hypothesis on $\langle\langle c\rangle\rangle(x - y)$ says that the quadratic form $\langle\langle c\rangle\rangle x$ is hyperbolic, so by [**EL 73**, 2.2], x is isomorphic to a sum $\oplus_{i=1}^r \langle c_i\rangle\langle\langle n_i\rangle\rangle$ such that $c_i \in k^\times$ and n_i

is a norm from $k(\sqrt{c})$. Then by Example 19.7 and equation (19.3), $\langle\langle c\rangle\rangle\lambda^2(x - r\mathcal{H})$ maps to

$$r\langle\langle c\rangle\rangle + \langle\langle c\rangle\rangle \sum_{i=1}^{r} \langle -n_i\rangle + \sum_{1\le i<j\le r} \langle\langle c, n_i, n_j\rangle\rangle \quad \text{in } W. \tag{20.2}$$

Because the n_i's are norms, the middle term equals $-r\langle\langle c\rangle\rangle$ and each of the forms $\langle\langle c, n_i, n_j\rangle\rangle$ is hyperbolic. That is, (20.2) is zero. □

20.3. PROPOSITION. *Let x be an even-dimensional quadratic form. The class of $\langle\langle c\rangle\rangle\lambda^2\hat{x}$ in the Witt ring depends only on the class of $\langle\langle c\rangle\rangle x$ in the Witt ring* (and not on c or x).

(See (19.9) for a definition of $\hat{x}$.)

PROOF. Write $\dim x = 2r$. Suppose that $\langle\langle c\rangle\rangle x$ equals $\langle\langle d\rangle\rangle y$ for some $d \in k^\times$ and $2s$-dimensional form y. We must show that $\langle\langle c\rangle\rangle\lambda^2\hat{x}$ and $\langle\langle d\rangle\rangle\lambda^2\hat{y}$ have the same image in the Witt ring. Since $\hat{x}$ and $\hat{y}$ only depend on x and y up to Witt-equivalence, we may add hyperbolic planes to x or y, and so assume that s equals r.

Case 1 ($c = d$): Suppose first that $\langle\langle c\rangle\rangle = \langle\langle d\rangle\rangle$ in the Witt ring, i.e., c and d are equal in $k^\times/k^{\times 2}$. We want to show that

$$\langle\langle c\rangle\rangle\lambda^2(x - r\mathcal{H}) = \langle\langle c\rangle\rangle\lambda^2(y - r\mathcal{H}) \quad \text{in } W. \tag{20.4}$$

In view of Example 19.7, it suffices to prove that $\langle\langle c\rangle\rangle(\lambda^2 x - \lambda^2 y)$ is zero in the Witt ring. Applying (19.5), we find:

$$\langle\langle c\rangle\rangle(\lambda^2 x - \lambda^2 y) = \langle\langle c\rangle\rangle(\lambda^2(x-y) + (x-y)y) \quad \text{in } \widehat{W}. \tag{20.5}$$

Since $\langle\langle c\rangle\rangle(x-y)$ is zero in the Witt ring, Lemma 20.1 gives that $\langle\langle c\rangle\rangle\lambda^2(x-y)$ is hyperbolic. We conclude that (20.5) is zero and hence that (20.4) holds.

Case 2 ($x = y$): Suppose now that $\langle\langle c\rangle\rangle x = \langle\langle d\rangle\rangle x$ for some $d \in k^\times$. We want to show that

$$\langle\langle c\rangle\rangle\lambda^2(x - r\mathcal{H}) = \langle\langle d\rangle\rangle\lambda^2(x - r\mathcal{H}) \quad \text{in } W. \tag{20.6}$$

Since

$$\langle\langle c\rangle\rangle - \langle\langle d\rangle\rangle = \langle -c, d\rangle = \langle d\rangle\langle\langle cd\rangle\rangle \quad \text{in } W$$

and $\langle\langle c\rangle\rangle x$ is isomorphic to $\langle\langle d\rangle\rangle x$, the form $\langle\langle cd\rangle\rangle x$ is hyperbolic. Applying Lemma 20.1, we find that $\langle\langle cd\rangle\rangle\lambda^2(x - r\mathcal{H})$ is zero in the Witt ring. That is, (20.6) holds.

Case 3 (general case): We now prove the full proposition. By Case 1, we may assume that c and d are distinct in $k^\times/k^{\times 2}$ and so apply Cor. B.5 to find an even-dimensional form τ such that

$$\langle\langle c\rangle\rangle x = \langle\langle c\rangle\rangle\tau = \langle\langle d\rangle\rangle\tau = \langle\langle d\rangle\rangle y.$$

Combining Cases 1 and 2 completes the proof. □

20.7. COROLLARY. *For every $c \in k^\times$ and $x \in I^n$ with $n \ge 1$, the class of $\langle\langle c\rangle\rangle P_n(x) \in I^{2n+1}$ depends only on the class of $\langle\langle c\rangle\rangle x$ in the Witt ring* (and not on c or x).

PROOF. By definition,

$$\langle\langle c\rangle\rangle P_n(x) = \langle\langle c\rangle\rangle\lambda^2\hat{x} - 2^{n-1}\langle\langle c\rangle\rangle x.$$

The first term on the right side depends only on the class of $\langle\langle c\rangle\rangle x$ by Prop. 20.3. □

We now construct invariants a_5 and a_6 of Spin_{12} as in Rost's paper [**Rost 99c**].

20.8. DEFINITION OF a_5. For $\eta \in H^1(k, \mathrm{Spin}_{12})$, Th. 17.13 says that the corresponding quadratic form $q_\eta \in H^1(k, \mathrm{SO}_{12})$ is isomorphic to $\langle\langle c\rangle\rangle x$ for some $c \in k^\times$ and 6-dimensional form $x \in I^2$. We define $a_5(\eta) \in H^5(k, \mathbb{Z}/2\mathbb{Z})$ to be $e_5(\langle\langle c\rangle\rangle P_2(x))$, equivalently, $(c)\cdot e_4(P_2(x))$; this is well defined by Cor. 20.7.

20.9. EXAMPLE. The invariant a_5 is not zero. Indeed, by Theorem 17.13, for $\eta \in H^1(k, \mathrm{Spin}_{12})$, the form q_η is equal to $\langle d\rangle\langle\langle c\rangle\rangle(\phi_1 - \phi_2)$ for some $c, d \in k^\times$ and 2-Pfister forms ϕ_1, ϕ_2. We compute:

$$\begin{aligned} a_5(\eta) &= (c)\cdot P_2(\langle d\rangle\phi_1 + \langle -d\rangle\phi_2) \\ &= (c)\cdot[(-1)\cdot(d)\cdot e_2(\phi_1) + (-1)\cdot(-d)\cdot e_2(\phi_2) + e_2(\phi_1)\cdot e_2(\phi_2)]. \end{aligned} \tag{20.10}$$

Replacing k by a purely transcendental extension of degree 6, it is easy to pick a 1-cocycle η such that $a_5(\eta)$ is not only nonzero, but has *symbol length* 3. That is, such that $a_5(\eta)$ can be written as a sum of 3 symbols in $H^3(k, \mathbb{Z}/2\mathbb{Z})$ and no fewer.

If -1 is a square in k, the formula (20.10) becomes

$$a_5(\eta) = (c)\cdot e_2(\phi_1)\cdot e_2(\phi_2). \tag{20.11}$$

20.12. LEMMA. *If q_η is isotropic* or *-1 is a square in k, then $a_5(\eta)$ is a symbol. If q_η is isotropic* and *-1 is a square in k, then $a_5(\eta)$ is zero.*

PROOF. If q_η is isotropic, then it is Witt-equivalent to a 10-dimensional form in I^3, hence by Example 17.8 it is isomorphic to $\langle d\rangle\langle\langle c\rangle\rangle\phi + 2\mathcal{H}$ for some $c, d \in k^\times$ and 2-Pfister form ϕ, equivalently, is equal to $\langle d\rangle\langle\langle c\rangle\rangle(\phi - 2\mathcal{H})$ in the Witt ring. In terms of Example 20.9, this says that ϕ_2 can be chosen to be hyperbolic, hence

$$a_5(\eta) = (c)\cdot(-1)\cdot(d)\cdot e_2(\phi_1).$$

The lemma follows from this observation. □

20.13. DEFINITION OF a_6. From here until the end of this section, *we assume that -1 is a square in k_0*; we remind the reader of this asumption by inserting "$\sqrt{-1} \in k_0$" in parentheses in the statements of results.

Prop. 10.2 applied to a_5 gives an invariant a_6 of Spin_{12} defined by setting

$$a_6(\eta) = a_5(\eta)\cdot(\alpha),$$

where α is a nonzero element of $k^\times$ represented by q_η.

Let k be the field obtained by adjoining indeterminates c, d, v_1, v_2, w_1, w_2. Let ϕ_i be the form $\langle\langle v_i, w_i\rangle\rangle$ and let $\eta \in H^1(k, \mathrm{Spin}_{12})$ be such that q_η is $\langle d\rangle\langle\langle c\rangle\rangle(\phi_1' - \phi_2')$. Then q_η represents dv_2 and by (20.11) we find:

$$a_6(\eta) = a_5(\eta)\cdot(dv_2) = (c)\cdot(v_1)\cdot(w_1)\cdot(v_2)\cdot(w_2)\cdot(d) \neq 0.$$

That is, a_6 is not the zero invariant.

20.14. PROPOSITION ($\sqrt{-1} \in k_0$). $\mathrm{Inv}_{k_0}^{\mathrm{norm}}(\mathrm{Spin}_{12}, \mathbb{Z}/2\mathbb{Z})$ *is a free $R_2(k_0)$-module with basis e_3 (the Rost invariant), a_5, a_6.*

PROOF. Under the hypothesis that -1 is a square in k_0, Example 17.12 gives a homomorphism $\mathrm{SO}(6) \times \boldsymbol{\mu}_4 \to \mathrm{Spin}_{12}$ that induces a surjection on H^1's such that the image of a pair $(q, c) \in H^1(k, \mathrm{SO}(6) \times \boldsymbol{\mu}_4)$ in $H^1(k, \mathrm{SO}_{12})$ is the quadratic form $\langle\langle c\rangle\rangle q$.

Recall from S20.6 or Example 19.14 that $\mathrm{Inv}_{k_0}^{\mathrm{norm}}(\mathrm{SO}(6), \mathbb{Z}/2\mathbb{Z})$ is a free $R_2(k_0)$-module with basis w_2, w_4, b. The image of $\mathrm{SO}(6)$ in Spin_{12} sits in a copy of SL_6, so

$H^1(*, \mathrm{SO}(6)) \to H^1(*, \mathrm{Spin}_{12})$ is the zero map. Applying Lemma 6.7 with $G = \boldsymbol{\mu}_4$ and $G' = \mathrm{SO}(6)$, we find that restricting invariants of Spin_{12} to $\mathrm{SO}(6) \times \boldsymbol{\mu}_4$ identifies $\mathrm{Inv}^{\mathrm{norm}}_{k_0}(\mathrm{Spin}_{12}, \mathbb{Z}/2\mathbb{Z})$ with a submodule of the free $R_2(k_0)$-module with basis

$$1 \cdot \underline{s}, \quad w_2 \cdot \underline{s}, \quad w_4 \cdot \underline{s}, \quad b \cdot \underline{s}.$$

By Example 19.14, the last three are restrictions of the invariants e_3, a_5, and a_6 of Spin_{12}, respectively. However, $\lambda \cdot 1 \cdot \underline{s}$ is not such a restriction for any nonzero $\lambda \in R_2(k_0)$. To see this, one argues as in 18.9, comparing the images of the trivial class and an indeterminate $(t) \in H^1(k_0(t), \boldsymbol{\mu}_4)$ in $H^1(k_0(t), \mathrm{Spin}_{12})$. □

20.15. Invariants of Spin_{11}. There are two invariants of Spin_{11} with values in $\mathbb{Z}/2\mathbb{Z}$ that we can find without doing any work. As always, one has the Rost/Arason invariant $e_3 \colon H^1(*, \mathrm{Spin}_{11}) \to H^3(*, \mathbb{Z}/2\mathbb{Z})$. On the other hand, the inclusion of Spin_{11} in Spin_{12} from Example 17.14 leads to an invariant of Spin_{11} of degree 5 via the composition

$$H^1(*, \mathrm{Spin}_{11}) \to H^1(*, \mathrm{Spin}_{12}) \xrightarrow{a_5} H^5(*, \mathbb{Z}/2\mathbb{Z}).$$

We denote this invariant also by a_5. (Note that restricting a_6 to Spin_{11} gives the zero invariant. Indeed, the image of $H^1(*, \mathrm{Spin}_{11})$ in $H^1(*, \mathrm{SO}_{12})$ consists of those forms that represent 1.)

Proposition ($\sqrt{-1} \in k_0$). *$\mathrm{Inv}^{\mathrm{norm}}_{k_0}(\mathrm{Spin}_{12}, \mathbb{Z}/2\mathbb{Z})$ is a free $R_2(k_0)$-module with basis e_3, a_5.*

Proof. As in the proof of Prop. 20.14, we restrict the invariants of Spin_{11} to the subgroup $\mathrm{SO}(5) \times \boldsymbol{\mu}_4$. Recall from S19.1 that $\mathrm{Inv}_{k_0}(\mathrm{SO}(5), \mathbb{Z}/2\mathbb{Z})$ is a free $\mathbb{Z}/2\mathbb{Z}$-module with basis $1, w_2, w_4$. Therefore, the set of normalized invariants of Spin_{11} with values in $\mathbb{Z}/2\mathbb{Z}$ is identified with a subspace of the free $R_2(k_0)$-module with basis

$$1 \cdot \underline{s}, \quad w_2 \cdot \underline{s}, \quad w_4 \cdot \underline{s}.$$

We have a commutative diagram

$$\begin{array}{ccc} H^1(k, \mathrm{SO}(5)) \times H^1(k, \boldsymbol{\mu}_4) & \longrightarrow & H^1(k, \mathrm{SO}(6)) \times H^1(k, \boldsymbol{\mu}_4) \\ \downarrow & & \downarrow \\ H^1(k, \mathrm{Spin}_{11}) & \longrightarrow & H^1(k, \mathrm{Spin}_{12}) \end{array}$$

The inclusion $\mathrm{SO}(5) \to \mathrm{SO}(6)$ is given by $g \mapsto \left(\begin{smallmatrix} g & 0 \\ 0 & 1 \end{smallmatrix}\right)$, so the arrow $H^1(k, \mathrm{SO}(5)) \to H^1(k, \mathrm{SO}(6))$ sends a 5-dimensional quadratic form q to $q \oplus \langle 1 \rangle$. The restriction of $w_j \colon H^1(k, \mathrm{SO}(6)) \to H^j(k, \mathbb{Z}/2\mathbb{Z})$ to $\mathrm{SO}(5)$ is

$$w_j(q \oplus \langle 1 \rangle) = (1) \cdot w_{j-1}(q) + w_j(q) = w_j(q),$$

so the invariants e_3 and a_5 of Spin_{12} restrict to $w_2 \cdot \underline{s}$ and $w_4 \cdot \underline{s}$ on $H^1(k, \mathrm{SO}(5) \times \boldsymbol{\mu}_4)$.

As in the proof of Prop. 20.14, one checks that $\lambda \cdot 1 \cdot \underline{s}$ is not the restriction of an invariant of Spin_{11} for any nonzero $\lambda \in R_2(k_0)$. □

21. Surjectivities: Spin_{14}

For this section, we fix a primitive 4-th root of unity i in a separable closure of k_0.

21.1. EXAMPLE ($(G_2 \times G_2) \rtimes \boldsymbol{\mu}_8 \subset \mathrm{Spin}_{14}$). Returning to the internal Chevalley modules defined in 9.11, we take $\widetilde{G}$ to be the split group of type E_8 and we omit the root $\pi := \alpha_1$. The semisimple subgroup G is simply connected of type D_7—i.e., it is isomorphic to Spin_{14}—and the representation V is a half-spin representation.

Fix a 7-dimensional quadratic form q such that $\langle 1 \rangle \oplus q$ is hyperbolic. We view Spin_{14} as the spin group of the quadratic form $q \oplus -q$, which gives a homomorphism $\mathrm{Spin}(q) \times \mathrm{Spin}(-q) \to \mathrm{Spin}_{14}$. We may identify the vector spaces underlying the form q and underlying the 7-dimensional fundamental representation of G_2 (which we call the standard representation of G_2) so that q is G_2-invariant. (Note that the standard representation of G_2 is irreducible since the characteristic is different from 2 [**GS**, p. 413].) This gives an embedding of G_2 in $\mathrm{Spin}(q)$, hence of $G_2 \times G_2$ in Spin_{14}.

We now argue as in Example 9.12. The restriction of the representation V to $\mathrm{Spin}(q) \times \mathrm{Spin}(-q)$ is the tensor product of the (8-dimensional) spin representations of $\mathrm{Spin}(q)$ and $\mathrm{Spin}(-q)$ [**McKP**, p. 291]. As in Example 17.5, each of these restricts to be a direct sum of the 7-dimensional irreducible representation of G_2 and a 1-dimensional trivial representation. We take v to be a tensor product of nonzero vectors that are fixed by the two G_2's.

The maximal proper parabolic subgroups of Spin_{14} have semisimple part that is simply connected [**SS**, 5.4b] of type

$$D_6, \quad A_2 \times D_5, \quad A_3 \times A_3, \quad A_1 \times A_1 \times A_4, \quad \text{or} \quad A_7,$$

and such a semisimple group has a representation with finite kernel and dimension ≤ 13. Therefore, $G_2 \times G_2$ — which has no representations with finite kernel of dimension < 14 [**Li**, 2.10] — cannot be contained in a proper parabolic of Spin_{14}. We conclude that $G_2 \times G_2$ is the identity component of the stabilizer N of $[v]$ in Spin_{14}. (Popov [**Po**, p. 225, Prop. 11] gives a proof that $G_2 \times G_2$ is the identity component of N using concrete computations in the half-spin representation in the style of Igusa's paper [**Ig**].)

Rather than computing the full stabilizer N, we compute instead the normalizer of $G_2 \times G_2$ in Spin_{14}, which contains N. Write W for the 14-dimensional vector space underlying $q \oplus -q$. The image of $G_2 \times G_2$ in $GL(W)$ has normalizer

$$((G_2.\mathbb{G}_m) \times (G_2.\mathbb{G}_m)) \rtimes \mathbb{Z}/2\mathbb{Z}, \tag{21.2}$$

where the nonidentity element in $\mathbb{Z}/2\mathbb{Z}$ is the matrix $\left(\begin{smallmatrix} 0 & 1 \\ 1 & 0 \end{smallmatrix}\right)$. The normalizer of $G_2 \times G_2$ in $\mathrm{SO}(W)$ is the intersection of (21.2) with $\mathrm{SO}(W)$, namely $(G_2 \times G_2) \rtimes \boldsymbol{\mu}_4$, where a primitive 4-th root of unity i in $\boldsymbol{\mu}_4$ is identified with the matrix

$$\begin{pmatrix} 0 & i \\ i & 0 \end{pmatrix} \in \mathrm{SO}(W).$$

Fix orthogonal bases $\{x_j\}$ and $\{y_j\}$ of the two standard representation of G_2 in W such that $q(x_j) = -q(y_j) = \pm 1$ for all j. The element

$$s := \prod_{j=1}^{7} \frac{1 + i x_j y_j}{\sqrt{2}}$$

in the even Clifford algebra belongs to Spin_{14}, has order 8 since $s^2 = \prod i x_j y_j$, and maps to $\left(\begin{smallmatrix} 0 & i \\ i & 0 \end{smallmatrix}\right)$ in $\mathrm{SO}(W)$. Therefore, the normalizer of $G_2 \times G_2$ in Spin_{14} is $(G_2 \times G_2) \rtimes \boldsymbol{\mu}_8$, where the copy of $\boldsymbol{\mu}_8$ is generated by s.

Th. 9.11 says that *the inclusion*

$$(G_2 \times G_2) \rtimes \boldsymbol{\mu}_8 \to \mathrm{Spin}_{14}$$

induces a surjection on H^1's.

We now interpret this result in terms of quadratic forms. Fix a quadratic extension $K := k(\sqrt{d})$ of k. The *trace* $\mathrm{tr}_*(q)$ of a quadratic form q over K is a quadratic form over k of dimension $2\dim q$. It is defined by viewing the K-vector space V underlying q as a vector space over k and taking the bilinear form

$$V \times V \xrightarrow{\text{bilinearization of } q} K \xrightarrow{\mathrm{tr}_{K/k}} k.$$

In other words, $\mathrm{tr}_*(q)$ is the Scharlau transfer of q via the linear map $\mathrm{tr}_{K/k}$, see e.g. [**Lam**, §VII.1].

The goal of this section is to prove:

21.3. THEOREM. (Rost [**Rost 99b**]) *Every 14-dimensional form in I^3k is of (at least) one of the following two types:*

(1) $\langle a\rangle(\phi_1' - \phi_2')$ *for some $a \in k^\times$ and ϕ_1, ϕ_2 3-Pfister forms over k.*

(2) $\mathrm{tr}_*(\sqrt{d}\phi')$ *for some nonsquare $d \in k^\times$ and ϕ a 3-Pfister form over $k(\sqrt{d})$.*

[Here we have written $'$ for the pure part of a Pfister form, so for example ϕ equals $\langle 1\rangle \oplus \phi'$ in (2).]

Case (1) can be viewed as the special case of (2) where the quadratic étale algebra $k[x]/(x^2 - d)$ is $k \times k$.

We remark that a 14-dimensional form in I^3 is as in (1) if and only if it contains a subform similar to a 2-Pfister form, see [**HT**, 2.3] or [**IK**, 17.2]. The two papers just cited give concrete examples of 14-dimensional forms that cannot be written as in (1), see [**HT**, p. 211] and [**IK**, 17.3]. Izhboldin and Karpenko applied Th. 21.3 to give a concrete description of 8-dimensional forms in I^2 whose Clifford algebra has index 4, see [**IK**, 16.10].

21.4. We proceed slowly to the proof of Th. 21.3. First, we compute the trace of a 1-dimensional form. Directly from the definition, we find:

(21.5) *For $\ell \in K^\times$, the 2-dimensional quadratic form $\mathrm{tr}_*(\langle\sqrt{d}\ell\rangle)$ represents $\mathrm{tr}_{K/k}(\sqrt{d}\ell) \in k$ and has determinant $-N_{K/k}(\ell) \in k^\times/k^{\times 2}$.*

That is,

$$\mathrm{tr}_*(\langle\sqrt{d}\ell\rangle) \cong \begin{cases} \text{hyperbolic plane} & \text{if } \ell \in k^\times, \text{ i.e., if } \mathrm{tr}_{K/k}(\sqrt{d}\ell) = 0; \\ \langle \mathrm{tr}_{K/k}(\sqrt{d}\ell)\rangle\langle 1, -N_{K/k}(\ell)\rangle & \text{otherwise.} \end{cases}$$

To see that this isomorphism holds, it suffices by [**Lam**, I.5.1] to observe that the forms on either side of the isomorphism have the same determinant and represent $\mathrm{tr}_{K/k}(\sqrt{d}\ell)$, which follows from (21.5).

Next we compute a toy example.

21.6. EXAMPLE. Write V for the vector space k^2 endowed with the quadratic form $q\colon \binom{x}{y} \mapsto x^2 - y^2$. Map the group $(\boldsymbol{\mu}_2 \times \boldsymbol{\mu}_2) \rtimes \boldsymbol{\mu}_4$ into the orthogonal group $\mathrm{O}(q)$ of q by sending

$$(\varepsilon_1, \varepsilon_2, i^r) \mapsto \begin{pmatrix} \varepsilon_1 & 0 \\ 0 & \varepsilon_2 \end{pmatrix} \begin{pmatrix} 0 & i \\ i & 0 \end{pmatrix}^r$$

for $\varepsilon_1, \varepsilon_2 \in \{\pm 1\}$ and $r \in \mathbb{Z}$. The set $H^1(k, \mathrm{O}(q))$ classifies 2-dimensional quadratic forms over k and we ask: Given a class $\eta \in H^1(k, (\boldsymbol{\mu}_2 \times \boldsymbol{\mu}_2) \rtimes \boldsymbol{\mu}_4)$, what is the 2-dimensional quadratic form q_η deduced from it?

The quotient map $(\boldsymbol{\mu}_2 \times \boldsymbol{\mu}_2) \rtimes \boldsymbol{\mu}_4 \to \boldsymbol{\mu}_4$ sends η to an element $\overline{\eta} \in H^1(k, \boldsymbol{\mu}_4)$, i.e., some $dk^{\times 4} \in k^\times / k^{\times 4}$.

If d is a square in k, then η comes from $H^1(k, \boldsymbol{\mu}_2 \times \boldsymbol{\mu}_2 \times \boldsymbol{\mu}_2)$, i.e., η corresponds to a triple $(\alpha, \beta, \gamma) \in (k^\times / k^{\times 2})^{\times 3}$. The 2-dimensional k-subspace of $V \otimes_k k_{\mathrm{sep}}$ fixed by $\eta_\sigma \sigma$ for all σ in the Galois group of k is spanned by

$$\begin{pmatrix} \sqrt{\alpha\gamma} \\ 0 \end{pmatrix} \quad \text{and} \quad \begin{pmatrix} 0 \\ \sqrt{\beta\gamma} \end{pmatrix}.$$

The quadratic form q_η is the restriction of q to this subspace, i.e., q_η is isomorphic to $\langle \gamma \rangle \langle \alpha, -\beta \rangle$.

Suppose now that d is not a square in k. Fix a 4-th root δ of d such that $\overline{\eta}_\sigma \sigma(\delta) = \delta$. Note that

$$\overline{\eta}_\sigma \sigma(\delta^3) = \begin{cases} \delta^3 & \text{if } \sigma \text{ is the identity on } K \\ -\delta^3 & \text{otherwise.} \end{cases}$$

If we twist $\boldsymbol{\mu}_2 \times \boldsymbol{\mu}_2$ by $\overline{\eta}$, we find the transfer $R_{K/k}(\boldsymbol{\mu}_2)$ for $K := k(\sqrt{d})$. Moreover, η is in the image of the map

$$K^\times / K^{\times 2} = H^1(k, R_{K/k}(\boldsymbol{\mu}_2)) \xrightarrow{\sim} H^1(k, (\boldsymbol{\mu}_2 \times \boldsymbol{\mu}_2)_{\overline{\eta}}) \to H^1(k, (\boldsymbol{\mu}_2 \times \boldsymbol{\mu}_2) \rtimes \boldsymbol{\mu}_4),$$

i.e., η is the image of a class $\ell K^{\times 2} \in K^\times / K^{\times 2}$. Write $\overline{\ell}$ for the image of ℓ under the nonidentity k-automorphism of K and fix square roots $\sqrt{\ell}, \sqrt{\overline{\ell}} \in k_{\mathrm{sep}}$. Then η is the image of the 1-cocycle

$$\sigma \mapsto \begin{cases} (\sigma(\sqrt{\ell})^{-1}\sqrt{\ell}, \sigma(\sqrt{\overline{\ell}})^{-1}\sqrt{\overline{\ell}}) & \text{if } \sigma \text{ is the identity on } K \\ (\sigma(\sqrt{\overline{\ell}})^{-1}\sqrt{\ell}, \sigma(\sqrt{\ell})^{-1}\sqrt{\overline{\ell}}) & \text{otherwise} \end{cases}$$

with values in $(\boldsymbol{\mu}_2 \times \boldsymbol{\mu}_2)_{\overline{\eta}}$. By considering separately the cases where σ is and is not the identity on K, it is easy to check that $\eta_\sigma \sigma$ fixes the vectors

$$\begin{pmatrix} \delta\sqrt{\ell} \\ \delta\sqrt{\overline{\ell}} \end{pmatrix} \quad \text{and} \quad \begin{pmatrix} \delta^3\sqrt{\ell} \\ -\delta^3\sqrt{\overline{\ell}} \end{pmatrix}$$

in $V \otimes k_{\mathrm{sep}}$; the quadratic form q_η is the restriction of q to the subspace they span. The value of q on the first vector is

$$\delta^2(\ell - \overline{\ell}) = \mathrm{tr}_{K/k}(\sqrt{d}\ell).$$

The determinant of the restriction q_η of q to this subspace is

$$\det \begin{pmatrix} \delta^2(\ell - \overline{\ell}) & \delta^4(\ell + \overline{\ell}) \\ \delta^4(\ell + \overline{\ell}) & \delta^6(\ell - \overline{\ell}) \end{pmatrix} = -4d^2 N_{K/k}(\ell).$$

As in 21.4, q_η is $\mathrm{tr}_*(\langle \sqrt{d}\ell \rangle)$.

Sketch of proof of Theorem 21.3. By Example 21.1, it suffices to describe the quadratic form deduced from a $(G_2 \times G_2) \rtimes \boldsymbol{\mu}_4$-torsor, as that is the image of $(G_2 \times G_2) \rtimes \boldsymbol{\mu}_8$ in $\mathrm{SO}(W)$. Reasoning as in Example 9.1, one can reduce the descent computation to the case of a 2-dimensional quadratic form. This computation was done in Example 21.6. □

21.7. REMARK. Roughly speaking, the theorems about quadratic forms q of dimension 8, 10, 12, and 14 in 17.6, 17.8, 17.13, and 21.3 respectively have as a crucial hypothesis that the discriminant of q is trivial and one component of the even Clifford algebra of q is split. With this in mind, one can ask if these theorems can be generalized to central simple algebras with orthogonal involution (A, σ) of degree 8, 10, 12, or 14 such that σ has trivial discriminant and one component of the even Clifford algebra is split. For degrees 8 and 12, generalizations are given in [**KMRT**, 42.11] and [**GQ b**] respectively. For degrees 10 and 14, the hypothesis on the even Clifford algebra implies that A is split, i.e., the "more general" situation amounts to the cases treated in 17.8 and 21.3.

22. Invariants of Spin_{14}

In this section, we exhibit some invariants of Spin_{14} with $\mathbb{Z}/2\mathbb{Z}$ coefficients using results from §21. The results here are all derived from [**Rost 99c**].

22.1. We define an invariant a_6 of Spin_{14} to be the composition

$$a_6 \colon H^1(k, \mathrm{Spin}_{14}) \longrightarrow I^3 \xrightarrow{P_3} I^6 \xrightarrow{e_6} H^6(k, \mathbb{Z}/2\mathbb{Z}),$$

where P_3 and e_6 are the maps from §19 and Example 1.2.3 respectively.

We argue that a_6 is not the zero invariant. For a given base field k, define k_1 to be the field obtained by adjoining six indeterminates t_{rs} for $r = 1, 2$ and $s = 1, 2, 3$, and—if it is not already in k—a square root of -1. Put $\phi_r := \langle\langle t_{r1}, t_{r2}, t_{r3}\rangle\rangle$ and take $\eta \in H^1(k_1, \mathrm{Spin}_{14})$ to have corresponding quadratic form $q_\eta = \phi_1' - \phi_2'$. By (19.13), we have:

$$a_6(\eta) = e_3(\phi_1) \cdot e_3(\phi_2) = \prod_{r,s} (t_{rs}) \neq 0.$$

22.2. PROPOSITION. *Fix $\eta \in H^1(k, \mathrm{Spin}_{14})$ and write q_η for the quadratic form deduced from it.*

(1) *If q_η is isotropic, then $a_6(\eta)$ is a symbol.*
(2) *If q_η is isotropic and -1 is a square in k, then $a_6(\eta)$ is zero.*
(3) *If -1 and 3 are squares in k, then $a_6(\eta)$ is a symbol.*

PROOF. If q_η is isotropic, then it is Witt-equivalent to a 12-dimensional form in I^3. By Th. 17.13, q_η equals $\langle d\rangle\langle\langle c\rangle\rangle(\phi_1 - \phi_2)$ in the Witt ring, for some $c, d \in k^\times$ and 2-Pfister forms ϕ_1, ϕ_2. Equation (19.13) gives:

$$a_6(\eta) = (c) \cdot (c) \cdot e_2(\psi_1) \cdot e_2(\psi_2).$$

This proves (1). If additionally -1 is a square in k, then $(c) \cdot (c)$ is zero, proving (2).

We now prove (3). Computing in the Witt ring, $P_3(q_\eta)$ is $7 + \lambda^2 q_\eta$ by Example 19.7, which equals $\lambda^2 q_\eta + \langle 1\rangle$.

By Exercise 19.2, the "reduced" Killing form (as defined in [**GN**, §5]) of $\mathfrak{so}(q_\eta)$ is $\langle -1\rangle\lambda^2 q_\eta$. The Lie algebra $\mathfrak{so}(q_\eta)$ contains a subalgebra of type $G_2 \times G_2$ or the transfer of a G_2 from a quadratic extension by Th. 21.3. The Killing form on a Lie algebra of type G_2 associated with a 3-Pfister form ψ is $\langle -1, -3\rangle\psi'$—see e.g. S27.21—so it contains a 7-dimensional totally isotropic subspace. Hence the reduced Killing form of $\mathfrak{so}(q_\eta)$ contains a totally isotropic subspace of dimension

14. By the previous paragraph, the class of $P_3(q_\eta)$ in the Witt ring is represented by an anisotropic quadratic form of dimension at most

$$(\dim \lambda^2 q_\eta) + 1 - 28 = 64.$$

But $P_3(q_\eta)$ belongs to I^6, so it is similar to a 6-Pfister form [**Lam**, X.5.6]. □

22.3. Suppose now that -1 is a square in k_0. We define an invariant

$$a_7 \colon H^1(*, \mathrm{Spin}_{14}) \to H^7(*, \mathbb{Z}/2\mathbb{Z}) \quad \text{via } a_7(\eta) := a_6(\eta) \cdot (\alpha)$$

where α is any nonzero element of k represented by q_η. By Propositions 22.2.2 and 10.2, this is a well-defined invariant of Spin_{14}.

22.4. EXAMPLE. (Assuming $\sqrt{-1} \in k_0$.) Let $\eta \in H^1(k, \mathrm{Spin}_{14})$ be such that q_η equals $\langle c \rangle (\phi_1' - \phi_2')$ for some $c \in k^\times$ and 3-Pfister forms ϕ_1, ϕ_2. Write ϕ_1 as $\langle\langle \alpha_1, \alpha_2, \alpha_3 \rangle\rangle$. we have

$$a_7(\eta) = (-c\alpha_1) \cdot e_3(\phi_1) \cdot e_3(\phi_2)$$

by (19.13). But $(-\alpha_1) \cdot e_3(\phi_1)$ is zero, hence

$$a_7(\eta) = (c) \cdot e_3(\phi_1) \cdot e_3(\phi_2).$$

As in 22.1, it is easy to see that a_7 is not the zero invariant.

23. Partial summary of results

Surjectivities. Table 23a summarizes the surjectivites proved above. The restrictions on the characteristic listed in the table should not be taken seriously. They only reflect the availability of easy-to-cite results in the literature.

N	$\subset$ G	char k_0	Ref.
$\mathrm{Spin}_{2n-1} \times \boldsymbol{\mu}_2$	$\subset \mathrm{Spin}_{2n}$	$\neq 2$	17.1
$G_2 \times \boldsymbol{\mu}_2$	$\subset \mathrm{Spin}_7$	$\neq 2$	17.5
$G_2 \times \boldsymbol{\mu}_2 \times \boldsymbol{\mu}_2$	$\subset \mathrm{Spin}_8$	$\neq 2$	17.1 and 17.5
$\mathrm{Spin}_7 \times \boldsymbol{\mu}_2$	$\subset \mathrm{Spin}_9$	$\neq 2$	17.7
$G_2 \times \boldsymbol{\mu}_4$	$\subset \mathrm{Spin}_{10}$	$\neq 2$	17.8
$\mathrm{SO}_5 \times \boldsymbol{\mu}_4$	$\subset \mathrm{Spin}_{11}$	$\neq 2$	17.14
$\mathrm{SO}_6 \times \boldsymbol{\mu}_4$	$\subset \mathrm{Spin}_{12}$	$\neq 2$	17.12
$(G_2 \times G_2) \rtimes \boldsymbol{\mu}_8$	$\subset \mathrm{Spin}_{14}$	$\neq 2$	21.1
$F_4 \times \boldsymbol{\mu}_3$	$\subset E_6$	any	9.12
$E_6 \rtimes \boldsymbol{\mu}_4$	$\subset E_7$	$\neq 2$	12.13

TABLE 23A. Examples of inclusions for which the morphism $H^1_{\mathrm{fppf}}(*, N) \to H^1(*, G)$ is surjective

This table is obviously not exhaustive. We have only considered a short list of internal Chevalley modules; the recipe in 9.11 gives others. For example, taking $\widetilde{G}$ to be E_6, E_7, E_8 and $\pi = \alpha_2$, one finds that there is an open GL_n-orbit in $\wedge^3 k^n$ ("alternating trilinear forms") for $n = 6, 7, 8$, hence an open SL_n-orbit in $\mathbb{P}(\wedge^3 k^n)$. Alternatively, other examples where there is an open G-orbit in $\mathbb{P}(V)$ can be found by consulting the table at the end of [**PoV**] or the lists of prehomogeneous vector spaces in [**SK**].

Invariants and essential dimensions of Spin groups. Table 23b summarizes the results on invariants of Spin_n for $n \leq 14$. We remark that in the examples considered in S (O_n, SO_n, the symmetric group on n letters, ...), the description of the invariants depended in a regular way on n; clearly, that is not the case here.

n	$\mathrm{ed}(\mathrm{Spin}_n)$	$\mathrm{Inv}^{\mathrm{norm}}_{k_0}(\mathrm{Spin}_n, \mathbb{Z}/2\mathbb{Z})$ has basis with elements of degree	Restrictions on k_0?	Ref.
1	1	1		16.1
2–6	0	$\varnothing$		16.1
7	4	3, 4		18.6
8	5	3, 4, 4, 5		18.1
9	5	3, 4, 5	$\sqrt{-1} \in k_0$	18.9
10	4	3, 4		18.7
11	5	3, 5	$\sqrt{-1} \in k_0$	20.15
12	6	3, 5, 6	$\sqrt{-1} \in k_0$	20.14
13	6	?		[**Rost 99c**, §10]
14	7	?	$\sqrt{-1} \in k_0$	22.3

TABLE 23B. Invariants and essential dimension of Spin_n for $n \leq 14$

All statements are under the global hypothesis that the characteristic of k_0 is $\neq 2$.

The values for the essential dimension given in the table are easily deduced from various results in Part III. For example, the table claims that the essential dimension of Spin_7 is 4. Since Spin_7 has a nontrivial cohomological invariant of degree 4, the essential dimension is ≥ 4, cf. 5.8. (All lower bounds on essential dimension here are proved by constructing nonzero cohomological invariants. These bounds can also be obtained by less ad hoc means, see [**CS**].) On the other hand, the essential dimension of $G_2 \times \boldsymbol{\mu}_2$ is 4, so the surjectivity from Example 17.5 shows that the essential dimension is ≤ 4.

What of Spin_{13}, which we have not yet discussed? One knows that the essential dimension is at least 6 by [**CS**] or because the invariant a_6 of Spin_{14} restricts to be nonzero on Spin_{13}. One cannot get an upper bound by imitating the methods of §17 to get a surjectivity in Galois cohomology because the spin representation V does not have an open orbit in $\mathbb{P}(V)$. See [**Rost 99c**] for a proof that the essential dimension is at most 6.

The paper [**BRV**] gives striking bounds on the essential dimension of Spin_n for larger values of n.

23.1. OPEN PROBLEM. (Reichstein-Youssin [**RY**, p. 1047]) Let k_0 be an algebraically closed field of characteristic zero. Does there exist a nonzero invariant $H^1(*, \mathrm{Spin}_n) \to H^{\lfloor n/2 \rfloor + 1}(*, \mathbb{Z}/2\mathbb{Z})$ when $n \equiv 0, \pm 1 \mod 8$?

[For $n = 7, 8, 9$, one has the invariants described in Examples 18.6, 18.1, and 18.9 above.]

Appendices

A. Examples of anisotropic groups of type E_7

We use cohomological invariants to give examples of algebraic groups of type E_7 that are

- strongly inner (in the sense of [**Tits 90**]), meaning they are obtained by twisting the split group of type E_7 by a cocycle taking values in the simply connected cover;
- anisotropic; and
- defined over "prime-to-2 closed" fields or split by an extension of dimension 2.

A.1. GROUPS OF TYPE E_7. Write E_7 for the split simply connected group of that type over a field k. The Rost invariant r_{E_7} recalled in Example 1.2.4 maps

$$r_{E_7}\colon H^1(*, E_7) \to H^3(*, \mathbb{Z}/12\mathbb{Z}(2)),$$

see [**Mer**, pp. 150, 154]. (In this appendix, the group $H^3(k, \mathbb{Z}/n\mathbb{Z}(2))$ is as defined in [**Mer**]. If the characteristic of k does not divide n, then $H^3(k, \mathbb{Z}/n\mathbb{Z}(2))$ is $H^3(k, \boldsymbol{\mu}_n^{\otimes 2})$, as in the main body of the notes. In any case, it is n-torsion.) The group $H^3(k, \mathbb{Z}/12\mathbb{Z}(2))$ is 12-torsion, and its 4- and 3-torsion are identified with $H^3(k, \mathbb{Z}/4\mathbb{Z}(2))$ and $H^3(k, \mathbb{Z}/3\mathbb{Z}(2))$ respectively. We write r' for the composition of r_{E_7} with the projection of $H^3(k, \mathbb{Z}/12\mathbb{Z}(2))$ onto its 4-torsion, i.e.:

$$r'\colon H^1(k, E_7) \xrightarrow{r_{E_7}} H^3(k, \mathbb{Z}/12\mathbb{Z}(2)) \to H^3(k, \mathbb{Z}/4\mathbb{Z}(2)).$$

PROPOSITION. *Suppose that, for $\eta \in H^1(k, E_7)$, the twisted group $(E_7)_\eta$ is isotropic. Then $2r'(\eta) = 0$. If furthermore k has characteristic $\neq 2$, then $r'(\eta)$ has symbol length ≤ 2 in $H^3(k, \mathbb{Z}/2\mathbb{Z})$.*

The last sentence of the proposition warrants some comments. The natural map $H^3(k, \mathbb{Z}/2\mathbb{Z}(2)) \to H^3(k, \mathbb{Z}/4\mathbb{Z}(2))$ identifies $H^3(k, \mathbb{Z}/2\mathbb{Z}(2))$ with the 4-torsion in $H^3(k, \mathbb{Z}/4\mathbb{Z}(2))$, whence we may view $r'(\eta)$ as an element of $H^3(k, \mathbb{Z}/2\mathbb{Z}(2))$. As the characteristic is not 2, this group is just $H^3(k, \mathbb{Z}/2\mathbb{Z})$. As for the symbol length, recall that every element of $H^3(k, \mathbb{Z}/2\mathbb{Z})$ can be written as a sum of symbols. The *symbol length* of a class in $z \in H^3(k, \mathbb{Z}/2\mathbb{Z})$ is the smallest natural number n such that z can be written as a sum of n symbols.

PROOF OF PROPOSITION A.1. We consult the list of the possible Tits indexes of groups of type E_7 from [**Tits 66**, p. 59]. (See 2.3 in that paper for the definition of the Tits index.) In three of these indexes ($E_{7,1}^{48}$, $E_{7,2}^{31}$, and $E_{7,4}^{9}$), one of the summands of the semisimple anisotropic kernel is of the from $SL(Q)$ for some quaternion division algebra Q. However, $(E_7)_\eta$ has trivial Tits algebras, so by [**Tits 71**, p. 211] the semisimple anisotropic kernel cannot have such a summand. In the remaining cases, the vertex 1 or 7 is circled, where the vertices are numbered as in Table 9. We refer to these possibilities as cases 1 and 7 respectively. If both vertices are circled, we arbitrarily say we are in case 1.

Fix a maximal split torus T in the split group E_7. As E_7 is simply connected and all roots have the same length, the cocharacter group T_* is identified with the root lattice. In case c, write S for the image of the cocharacter corresponding to twice the fundamental weight ω_c. Write G for the derived subgroup of the centralizer of S in E_7; it is simply connected and split; it has type D_6 in case 1 and type E_6 in case 7. For precision, we write i for the inclusion $G \hookrightarrow E_7$ and i_* for the induced map on H^1's.

By Tits's Witt-type theorem, $(E_7)_\eta$ is isomorphic to $(E_7)_{i_*\tau}$ for some class τ in $H^1(k, G)$. It follows that $i_*\tau = \zeta \cdot \eta$ where ζ is a 1-cocycle taking values in the center Z of E_7. As the Rost invariant is compatible with twisting [**Gi 00**, p. 76, Lem. 7], we have

$$r'(i_*\tau) = r'(\zeta \cdot \eta) = r'(\zeta) + r'(\eta),$$

cf. [**Ga 01a**, 7.1]. However, E_7 is split, so the image of $H^1(k, Z)$ in $H^1(k, E_7)$ is zero. In particular, $r'(\zeta)$ is zero and $r'(i_*\tau) = r'(\eta)$. Replacing η with $i_*\tau$, we may assume that η is the image of τ. Since the inclusion i of G in E_7 has Rost multiplier one, $r_{E_7}(\eta)$ equals $r_G(\tau)$.

To prove the first claim in the proposition, it suffices to observe that the order of r_G is 2 in case 1 and 6 in case 7 by [**Mer**, 15.4, 16.6]. In both cases, the 2-primary part is 2 and not 4.

We now prove the last claim. In case 7, the "mod 2" portion of the Rost invariant is a symbol over an odd-dimension extension of k by S22.9, hence it is a symbol over k by [**Rost 99a**], see Example 8.9. In case 1, G is the split group Spin_{12} and by Th. 17.13 the quadratic form $q_\tau \in H^1(k, \mathrm{SO}_{12})$ deduced from τ is of the form $q_\tau = \langle d \rangle \langle\langle c \rangle\rangle (\phi_1' - \phi_2')$ for some $c, d \in k^\times$ and 2-Pfister forms ϕ_1, ϕ_2. The Rost invariant of τ is

$$r_G(\tau) = e_3(q_\tau) = (c) \cdot e_2(\phi_1) + (c) \cdot e_2(\phi_2),$$

a sum of two symbols. □

Suppose that k is a prime-to-2 closed field, i.e., every finite extension of k has dimension a power of 2. Every group of inner type E_6 is isotropic. In case $k = \mathbb{R}$, the unique anisotropic simply connected group of type E_7 is not a strongly inner form of E_7 (i.e., it has nontrivial Tits algebras). We now give an example of a prime-to-2 closed field k that supports an anisotropic strongly inner form of E_7.

A.2. Example. Rost [**Gi 00**, p. 91, Prop. 8] gives an extension k_0 of $\mathbb{Q}$ and a class $\eta \in H^1(k_0, E_7)$ such that $2r'(\eta)$ is not zero. If we take k to be the extension of k_0 fixed by a 2-Sylow subgroup of the absolute Galois group of k_0, then every finite extension of k has dimension a power of 2, yet k supports the strongly inner form of E_7 obtained by twisting by $\mathrm{res}_{k/k_0}(\eta)$, and this group is anisotropic by the proposition.

In the preceding example, $2r'(\eta)$ is not zero over k, so a restriction/corestriction argument shows that the twisted group $(E_7)_\eta$ is not split by a quadratic extension of k. We can use the second criterion in the proposition to give an example of a strongly inner form of E_7 that is anisotropic but is split by a quadratic extension.

A.3. Example. Let F be a field of characteristic zero such that -1 is a square in F. Let k_0 be the field obtained by adjoining the indeterminates $t_1, t_2, \ldots, t_6$ and put k for the field $k_0(d)$, for d an indeterminate. We construct a strongly inner form G of E_7 that is anisotropic over k and split over the quadratic extension $K := k(\sqrt{d})$. Let H denote the quasi-split simply connected group of type ${}^2E_6$ associated with the quadratic extension K/k; it is a subgroup of the split simply connected group E_7 and the inclusion has Rost multiplier 1. Chernousov [**Ch 03**, p. 321] gives a 1-cocycle $\eta \in H^1(K/k, H)$ whose image under r' is

$$(d) \cdot [(t_1) \cdot (t_3) + (t_2 t_3 t_5) \cdot (t_4) + (t_5) \cdot (t_6)] \quad \in H^3(k, \mathbb{Z}/2\mathbb{Z}). \tag{A.4}$$

We take G to be E_7 twisted by η. As η is killed by K, G is K-split. For sake of contradiction, suppose that G is isotropic over k. Applying Prop. A.1, we note that $r'(\eta)$ can be written as a sum of ≤ 2 symbols in $H^3(k, \mathbb{Z}/2\mathbb{Z})$. It follows that the residue with respect to d, namely

$$(t_1)\cdot(t_3) + (t_2t_3t_5)\cdot(t_4) + (t_5)\cdot(t_6) \quad \in H^2(k_0, \mathbb{Z}/2\mathbb{Z}) \tag{A.5}$$

can be written as a sum of ≤ 2 symbols in $H^2(k_0, \mathbb{Z}/2\mathbb{Z})$. For P_n as in Prop. 19.12.3, the image of (A.5) under the composition

$$H^2(k_0, \mathbb{Z}/2\mathbb{Z}) \xrightarrow[e_2^{-1}]{\sim} I^2/I^3 \xrightarrow{P_2} I^4/I^5 \xrightarrow[e_4]{\sim} H^4(k_0, \mathbb{Z}/2\mathbb{Z})$$

is

$$(t_1)\cdot(t_3)\cdot(t_2t_5)\cdot(t_4) + (t_1)\cdot(t_3)\cdot(t_5)\cdot(t_6) + (t_2t_3)\cdot(t_4)\cdot(t_5)\cdot(t_6).$$

By parts 1 and 4 of Prop. 19.12, this is a (possibly zero) symbol in $H^4(k_0, \mathbb{Z}/2\mathbb{Z})$. Taking residues with respect to t_2 and then t_4, we find

$$(t_1)\cdot(t_3) + (t_5)\cdot(t_6) \quad \in H^3(F(t_1, t_3, t_5, t_6), \mathbb{Z}/2\mathbb{Z}).$$

Our assumption implies that this is a symbol, which is impossible as the t_i's are indeterminates. We conclude that G is anisotropic over k.

A.6. We close with an application to groups of type E_8. Let G be an anisotropic simply connected group of type E_7 over a field k as constructed in Examples A.2 or A.3. In particular, G is anisotropic and has trivial Tits algebras. Then there is a group of type E_8 over k with Tits index

and semisimple anisotropic kernel G. (This follows from Tits's Witt-type theorem as in [**Tits 90**, p. 658, Cor. 1].) The existence of groups with this index was a question mark in [**Tits 66**, p. 60] that was settled in [**Tits 90**, p. 664, Prop. 2B].

B. A generalization of the Common Slot Theorem
By Detlev W. Hoffmann

The purpose of this appendix is to prove Cor. B.5, which is used in the construction of the degree 5 invariant of Spin_{12} in §20. The corollary as such is due to Rost, but his original argument had a small flaw. The version we present here is actually more general and can be considered as a generalization of the well known Common Slot Theorem, see, e.g., [**Lam**, III.4.13]. Recall that the Common Slot Theorem says that if $A = \left(\frac{a,x}{k}\right)$ and $B = \left(\frac{b,y}{k}\right)$ are quaternion algebras over a field k with $\mathrm{char}(k) \neq 2$ such that $A \cong B$, then there exists $z \in k^*$ with $A \cong \left(\frac{a,z}{k}\right)$ and $B \cong \left(\frac{b,z}{k}\right)$. Translated into Pfister forms, it means that if $\langle\langle a, x\rangle\rangle \cong \langle\langle b, y\rangle\rangle$ then

$$\langle\langle a, x\rangle\rangle \cong \langle\langle a, z\rangle\rangle \cong \langle\langle b, z\rangle\rangle \cong \langle\langle b, y\rangle\rangle$$

for some $z \in k^*$. Furthermore, $z \in D_k(\langle\langle ab\rangle\rangle)$, i.e., z is represented by the form $\langle\langle ab\rangle\rangle$ and hence is a norm of the extension $k(\sqrt{ab})/k$. Indeed, $\langle 1, -a, -z, az\rangle \cong \langle\langle a, z\rangle\rangle \cong \langle\langle b, z\rangle\rangle \cong \langle 1, -b, -z, bz\rangle$ implies after Witt cancellation and scaling that $\langle\langle ab\rangle\rangle \cong \langle z\rangle\langle\langle ab\rangle\rangle$.

In the sequel, all fields are assumed to be of characteristic different from 2. To state our version, we first recall the notion of *linkage* of Pfister forms introdued by Elman and Lam [**EL 72**]. Let α and β be Pfister forms over k of folds m and n, respectively. Then α and β are called *r-linked* for some nonnegative integer $r \leq \min(n, m)$ if there exist Pfister forms ρ, σ, τ of folds $m - r$, $n - r$ and r, respectively, such that $\alpha \cong \rho\tau$ and $\beta \cong \sigma\tau$. In other words, α and β are r-linked if they can be written with r slots in common. It can be shown that α and β are r-linked if and only if the Witt index of $\alpha \oplus \langle -1\rangle\beta$ is $\geq 2^r$ (see [**EL 72**, 4.4]).

If $m \geq n$, we call α and β *linked* if they are $(n-1)$-linked in the above sense, and we say that they are *strictly linked* if they are $(n-1)$-linked but not n-linked (i.e., α is not similar to a subform of β). So if $n = m$, being (strictly) linked means that there exist an $(n-1)$-fold Pfister form π and $a, b \in k^*$ such that $\alpha \cong \langle\langle a\rangle\rangle\pi$ and $\beta \cong \langle\langle b\rangle\rangle\pi$ (and $\alpha \ncong \beta$). Note that in this situation, we have in the Witt ring Wk that $\alpha - \beta = \langle b\rangle\langle\langle ab\rangle\rangle\pi$.

Recall also that a form ϕ is called *round* if $\phi \cong \langle a\rangle\phi$ if and only if $a \in D_k(\phi)$, i.e., the group of similarity factors $G_k(\phi)$ coincides with the set $D_k(\phi)$ of nonzero elements represented by ϕ. It is well known that Pfister forms are round. The following facts about round forms are also well known, see [**WS**, Theorem 2] for a proof (or [**EL 72**, 1.4] in case of Pfister forms).

B.1. LEMMA. *Let α and q be forms over k and assume that α is round.*

(1) *If $x \in D_k(\alpha q)$, then there exists a form q_1 such that $\alpha q \cong \alpha(\langle x\rangle \oplus q_1)$.*
(2) *If α is anisotropic and αq isotropic, then there exists a form q_2 such that $\alpha q \cong \alpha(\mathcal{H} \oplus q_2)$.*

The crucial ingredient in the proof of our result is the following theorem by Wadsworth and Shapiro [**WS**, Theorem 3].

B.2. THEOREM. *Let α and β be strictly linked Pfister forms over k of folds m and n, respectively, with $m \geq n \geq 1$. Let q be an anisotropic form over k and suppose that there exist forms ϕ and ψ over k with $q \cong \alpha\phi \cong \beta\psi$. Then there exist forms q_i, ϕ_i, ψ_i, $1 \leq i \leq r$, such that*

- *$q \cong q_1 \oplus q_2 \oplus \cdots \oplus q_r$, and*
- *$\dim \phi_i = 2$, $\dim \psi_i = 2^{m-n+1}$ for each i, and*
- *$q_i \cong \alpha\phi_i \cong \beta\psi_i$ for each i.*

Our result now reads as follows.

B.3. Proposition. *Let α and β be n-fold Pfister forms over k that are strictly linked. Let π be an $(n-1)$-fold Pfister form and $a, b \in k^*$ such that $\alpha \cong \pi\langle\langle a\rangle\rangle$ and $\beta \cong \pi\langle\langle b\rangle\rangle$, and let $\gamma \cong \pi\langle\langle ab\rangle\rangle$.*

If ϕ, ψ are forms over k such that $\alpha\phi = \beta\psi$ in Wk, then there exists a form τ over k, a nonnegative integer r, $c_i \in k^*$ and $d_i \in D_k(\gamma)$ $(1 \leq i \leq r)$ such that $\tau \cong \bigoplus_{i=1}^r \langle c_i\rangle\langle\langle d_i\rangle\rangle$, $\alpha\tau$ anisotropic and

$$\alpha\phi = \alpha\tau = \beta\tau = \beta\psi \in Wk\ .$$

Proof. Note that the assumption on α and β being strictly linked implies that γ is anisotropic.

By Lemma B.1(2), we may assume that $\alpha\phi$ and $\beta\psi$ are anisotropic and hence $\alpha\phi \cong \beta\psi$. We denote this anisotropic form by q and apply Theorem B.2 to deduce that there exist forms q_i, ϕ_i, ψ_i, $1 \leq i \leq r$ such that

- $q \cong q_1 \oplus q_2 \oplus \cdots \oplus q_r$, and
- $\dim \phi_i = \dim \psi_i = 2$ for each i, and
- $q_i \cong \alpha\phi_i \cong \beta\psi_i$ for each i.

But then, by Lemma B.1(1), there exist $c_i, x_i, y_i \in k^*$ such that $c_i \in D_k(q_i)$ and

$$q_i \cong \langle c_i\rangle\alpha\langle\langle x_i\rangle\rangle \cong \langle c_i\rangle\beta\langle\langle y_i\rangle\rangle\ .$$

Hence, $\alpha\langle\langle x_i\rangle\rangle \cong \beta\langle\langle y_i\rangle\rangle$, and with $\alpha \cong \pi\langle\langle a\rangle\rangle$, $\beta \cong \pi\langle\langle b\rangle\rangle$, $\gamma \cong \pi\langle\langle ab\rangle\rangle$, we get in Wk that

$$0 = \alpha\langle\langle x_i\rangle\rangle - \beta\langle\langle y_i\rangle\rangle = \alpha - \beta + \langle y_i\rangle\beta - \langle x_i\rangle\alpha$$

and therefore

$$\langle b\rangle\gamma = \langle x_i\rangle\alpha - \langle y_i\rangle\beta\ .$$

Comparing dimensions shows that $\langle x_i\rangle\alpha \oplus \langle -y_i\rangle\beta$ is isotropic. Thus, there exists $d_i \in D_k(\langle x_i\rangle\alpha) \cap D_k(\langle y_i\rangle\beta)$, and by Lemma B.1(1), we have $\langle x_i\rangle\alpha \cong \langle d_i\rangle\alpha$ and $\langle y_i\rangle\beta \cong \langle d_i\rangle\beta$. We conclude that $\alpha\langle\langle x_i\rangle\rangle \cong \alpha\langle\langle d_i\rangle\rangle$ and $\beta\langle\langle y_i\rangle\rangle \cong \beta\langle\langle d_i\rangle\rangle$, hence

$$q_i \cong \langle c_i\rangle\langle\langle d_i\rangle\rangle\alpha \cong \langle c_i\rangle\langle\langle d_i\rangle\rangle\beta\ .$$

The proof is now finished by putting $\tau \cong \bigoplus_{i=1}^r \langle c_i\rangle\langle\langle d_i\rangle\rangle$. □

B.4. Remarks. (i) One could relax the condition on being strictly linked by linked, and the above statement would still hold provided $\dim\phi$ is even. But this doesn't really yield anything new of interest. Indeed, if α and β are linked but not strictly so, then this just means that $\alpha \cong \beta$ which in turn implies that γ is hyperbolic. Hence, $D_k(\gamma) = k^*$. By Lemma B.1, there exists a (necessarily even-dimensional) form τ such that $(\alpha\phi)_{\mathrm{an}} \cong \alpha\tau$, and one can simply take *any* orthogonal decomposition $\tau \cong \bigoplus_{i=1}^r \langle c_i\rangle\langle\langle d_i\rangle\rangle$.

(ii) Recall that a field k is called *linked* if any two Pfister forms over k are linked. In fact, it is not difficult to check that k is linked iff any two 2-fold Pfister forms over k are linked. This notion of a linked field has been coined in [**EL 73**]. Well known examples of linked fields are finite, local and global fields, fields of transcendence degree ≤ 1 over a real closed field or of transcendence degree ≤ 2 over an algebraically closed field.

Hence, if we assume the field k in Proposition B.3 to be linked, then the condition of the two Pfister forms being strictly linked can be replaced by the two Pfister forms being nonisometric.

Let us state the case $n = 1$ for the above proposition explicitly.

B.5. COROLLARY. *Let $a, b \in k^*$ represent different nontrivial square classes. Let $\ell = k(\sqrt{ab})$. If ϕ, ψ are forms over k such that $\langle\langle a\rangle\rangle\phi = \langle\langle b\rangle\rangle\psi$ in Wk, then there exists a form τ over k, a nonnegative integer r, $c_i \in k^*$ and $d_i \in D_k(\langle\langle ab\rangle\rangle) = N_{\ell/k}(\ell^*)$ $(1 \le i \le r)$ such that $\tau \cong \bigoplus_{i=1}^r \langle c_i\rangle\langle\langle d_i\rangle\rangle$, $\langle\langle a\rangle\rangle\tau$ anisotropic and*

$$\langle\langle a\rangle\rangle\phi = \langle\langle a\rangle\rangle\tau = \langle\langle b\rangle\rangle\tau = \langle\langle b\rangle\rangle\psi \in Wk\ .$$

Suppose that, as in the corollary, a and b represent different nontrivial square classes in k^*. Let ϕ be an anisotropic form over k. If $\langle\langle a\rangle\rangle\phi$ is isotropic, it is well known and not difficult to see that there exists a 2-dimensional subform ϕ' of ϕ such that already $\langle\langle a\rangle\rangle\phi'$ is isotropic (and hence hyperbolic as it is similar to a 2-fold Pfister form), cf. [**EL 73**, 2.2].

Indeed, $\langle\langle a\rangle\rangle\phi \cong \phi\oplus\langle -a\rangle\phi$ being isotropic clearly implies that there are nonzero vectors x, y in an underlying vector space V of ϕ such that $\phi(x) = a\phi(y)$. Since a is not a square, x and y span a 2-dimensional subspace W of V. Then just take ϕ' to be the restriction of ϕ to W.

Now let $K = k(\sqrt{b})$ and suppose that $\langle\langle a\rangle\rangle\phi_K$ is isotropic (or possibly even hyperbolic, in which case $\langle\langle a\rangle\rangle\phi \cong \langle\langle b\rangle\rangle\psi$ for some form ψ, see [**Lam**, VII.3.2]). By the above, we see that there exists over K (!) a 2-dimensional subform ϕ' of ϕ_K such that $\langle\langle a\rangle\rangle\phi'$ is isotropic over K. If, in this situation, one could always find a 2-dimensional subform ϕ' of ϕ already over k (!) such that $\langle\langle a\rangle\rangle\phi'_K$ is isotropic (and hence hyperbolic) over K, then one could use the Common Slot Theorem plus a staightforward induction on $\dim\phi$ to easily deduce the above corollary. In fact, for $\dim\phi = 2$, the above corollary is essentially nothing else but the Common Slot Theorem.

However, such a 2-dimensional subform ϕ' of ϕ over k doesn't exist in general as the following counterexamples will show for forms ϕ of dimension n for any given $n > 2$.

B.6. EXAMPLE. Recall that the Pythagoras number $p(k)$ of a field k is defined to be the least positive integer p (provided such an integer exists) such that each sum of squares in k can be written as a sum of $\le p$ squares. If no such integer exists, then we put $p(k) = \infty$.

Let k be a formally real field with $p(k) = \infty$ (e.g., the rational function field over the reals in infinitely many variables, cf. [**Lam**, IX.2.4]). Let $n \ge 3$ and s be such that $2^s < n \le 2^{s+1}$. Pick an element $-b$ that is a sum of $2^{s+1}+2$ squares but not fewer. Note that this is always possible since $p(k) > 2^{s+1}+1$. Now let $a = -1$, so $\langle\langle a\rangle\rangle \cong \langle 1, 1\rangle$, $\phi \cong \langle 1, \ldots, 1\rangle$ (sum of n squares), and let $K = k(\sqrt{b})$.

Then $\langle\langle a\rangle\rangle\phi$ is a Pfister neighbor of $P \cong \langle\langle -1, -1, \ldots, -1\rangle\rangle$, a sum of 2^{s+2} squares. Now $P \cong \langle 1\rangle \oplus P'$ with P' a sum of $2^{s+2}-1$ squares. In particular, P' represents $-b$, and P has therefore a subform $\langle 1, -b\rangle$ which becomes isotropic over K. Hence P_K is hyperbolic and the Pfister neighbor $\langle\langle a\rangle\rangle\phi_K$ is isotropic. Note that if $n = 2^{s+1}$ then in fact $\langle\langle a\rangle\rangle\phi_K \cong P_K$ is hyperbolic.

Suppose now that ϕ contains a subform $\langle u, v\rangle$ over k with $\langle\langle a\rangle\rangle\langle u, v\rangle \cong \langle 1, 1\rangle\langle u, v\rangle$ isotropic over K. Note that both u and v are necessarily sums of $n \le 2^{s+1}$ squares in k as both are represented by ϕ.

Let $w = uv$. Then $\langle 1,1,w,w\rangle \cong \langle\langle -1,-w\rangle\rangle$ is similar to $\langle 1,1\rangle\langle u,v\rangle$ and thus isotropic (and hence hyperbolic) over K. But then b can be chosen as a slot of the Pfister form $\langle\langle -1,-w\rangle\rangle$: $\langle\langle -1,-w\rangle\rangle \cong \langle\langle b,c\rangle\rangle$ for some $c \in k^*$ (cf. [**Lam**, III.4.1]). By Witt cancellation, $\langle 1,w,w\rangle \cong \langle -b,-c,bc\rangle$ and thus $-b$ is represented by $\langle 1,w,w\rangle$. In particular, there exist $x,y,z \in k^*$ with $-b = x^2 + w(y^2+z^2)$.

Now $w(y^2+z^2) = uv(y^2+z^2)$ is the product of three factors, each of which being a sum of at most 2^{s+1} squares. A famous result by Pfister states that, for each nonnegative integer m, the nonzero sums of 2^m squares in a field form a multiplicative group (see, e.g., [**Lam**, X.1.9]). Hence, we have that $w(y^2+z^2)$ can be expressed itself as a sum of at most 2^{s+1} squares. But then, $-b$ can be written as a sum of at most $2^{s+1}+1$ squares, a contradiction!

Bibliography

[ABS] H. Azad, M. Barry, and G. Seitz, *On the structure of parabolic subgroups*, Comm. Algebra **18** (1990), no. 2, 551–562.

[Alb] A.A. Albert, *Non-cyclic division algebras of degree and exponent four*, Trans. Amer. Math. Soc. **35** (1933), 112–121.

[AM] A. Adem and R.J. Milgram, *Cohomology of finite groups*, second ed., Grundlehren der Mathematischen Wissenschaften, vol. 309, Springer-Verlag, Berlin, 2004.

[Ba] K. Baur, *Richardson elements for classical Lie algebras*, J. Algebra **297** (2006), 168–185.

[Bh] M. Bhargava, *Higher composition laws I: a new view on Gauss composition and quadratic generalizations*, Ann. of Math. (2) **159** (2004), 217–250.

[Bor] A. Borel, *Linear algebraic groups*, second ed., Graduate Texts in Mathematics, vol. 126, Springer-Verlag, New York, 1991.

[Bou Alg] N. Bourbaki, *Algebra II: Chapters 4–7*, Springer-Verlag, Berlin, 1988.

[Bou Lie] ———, *Lie groups and Lie algebras: Chapters 4–6*, Springer-Verlag, Berlin, 2002.

[BR] J. Buhler and Z. Reichstein, *On the essential dimension of a finite group*, Compositio Math. **106** (1997), 159–179.

[BRV] P. Brosnan, Z. Reichstein, and A. Vistoli, *Essential dimension and algebraic stacks*, preprint, January 2007.

[BS] A. Borel and J-P. Serre, *Théorèmes de finitude en cohomologie galoisienne*, Comment. Math. Helv. **39** (1964), 111–164 (= Borel, Oe., vol. 2, #64).

[BT] A. Borel and J. Tits, *Groupes réductifs*, Inst. Hautes Études Sci. Publ. Math. **27** (1965), 55–150.

[CG] M. Carr and S. Garibaldi, *Geometries, the principle of duality, and algebraic groups*, Expo. Math. **24** (2006), 195–234.

[CGR] V. Chernousov, Ph. Gille, and Z. Reichstein, *Resolving G-torsors by abelian base extensions*, J. Algebra **296** (2006), 561–581.

[Ch 95] V. Chernousov, *A remark on the* $(\bmod 5)$*-invariant of Serre for groups of type* E_8, Math. Notes **56** (1995), no. 1-2, 730–733, [Russian original: Mat. Zametki **56** (1994), no. 1, pp. 116-121].

[Ch 03] ———, *The kernel of the Rost invariant, Serre's Conjecture II and the Hasse principle for quasi-split groups* ${}^{3,6}D_4$, E_6, E_7, Math. Ann. **326** (2003), 297–330.

[CS] V. Chernousov and J-P. Serre, *Lower bounds for essential dimensions via orthogonal representations*, J. Algebra **305** (2006), 1055–1070.

[DG 70] M. Demazure and P. Gabriel, *Groupes algébriques. Tome I: Géométrie algébrique, généralités, groupes commutatifs*, Masson, Paris, 1970.

[DG 02] P. Deligne and B.H. Gross, *On the exceptional series, and its descendants*, C. R. Math. Acad. Sci. Paris **335** (2002), no. 11, 877–881.

[Dr] P.K. Draxl, *Skew fields*, London Math. Soc. Lecture Note Series, vol. 81, Cambridge University Press, Cambridge-New York, 1983.

[Dy 57a] E.B. Dynkin, *Maximal subgroups of the classical groups*, Amer. Math. Soc. Transl. (2) **6** (1957), 245–378 [Russian original: Trudy Moskov. Mat. Obšč. **1** (1952), 39–166].

[Dy 57b] ———, *Semisimple subalgebras of semisimple Lie algebras*, Amer. Math. Soc. Transl. (2) **6** (1957), 111–244 [Russian original: Mat. Sbornik N.S. **30(72)** (1952), 349–462].

[EKM] R.S. Elman, N. Karpenko, and A. Merkurjev, *The algebraic and geometric theory of quadratic forms*, Amer. Math. Soc. Colloquium Publications, vol. 56, 2008.

[EL 72] R. Elman and T.Y. Lam, *Pfister forms and K-theory of fields*, J. Algebra **23** (1972), 181–213.

[EL 73] ———, *Quadratic forms and the u-invariant. I*, Math. Z. **131** (1973), 283–304.

[Fe] J.C. Ferrar, *Strictly regular elements in Freudenthal triple systems*, Trans. Amer. Math. Soc. **174** (1972), 313–331.

[Ga 98] R.S. Garibaldi, *Isotropic trialitarian algebraic groups*, J. Algebra **210** (1998), 385–418.

[Ga 01a] ———, *The Rost invariant has trivial kernel for quasi-split groups of low rank*, Comment. Math. Helv. **76** (2001), no. 4, 684–711.

[Ga 01b] ———, *Structurable algebras and groups of type E_6 and E_7*, J. Algebra **236** (2001), no. 2, 651–691.

[GH] S. Garibaldi and D.W. Hoffmann, *Totaro's question on zero-cycles on G_2, F_4, and E_6 torsors*, J. London Math. Soc. **73** (2006), 325–338.

[Gi 00] Ph. Gille, *Invariants cohomologiques de Rost en caractéristique positive*, K-Theory **21** (2000), 57–100.

[Gi 02a] ———, *Algèbres simples centrales de degré 5 et E_8*, Canad. Math. Bull. **45** (2002), no. 3, 388–398.

[Gi 02b] ———, *An invariant of elements of finite order in semisimple simply connected algebraic groups*, J. Group Theory **5** (2002), no. 2, 177–197.

[GMS] S. Garibaldi, A.S. Merkurjev, and J-P. Serre, *Cohomological invariants in Galois cohomology*, University Lecture Series, vol. 28, Amer. Math. Soc., 2003.

[GN] B.H. Gross and G. Nebe, *Globally maximal arithmetic groups*, J. Algebra **272** (2004), 625–642.

[GQ a] S. Garibaldi and A. Quéguiner-Mathieu, *Restricting the Rost invariant to the center*, St. Petersburg Math. J. **19** (2008), 197–213.

[GQ b] ———, *Pfister's theorem for orthogonal involutions of degree 12*, preprint, 2008.

[GS] P.B. Gilkey and G.M. Seitz, *Some representations of exceptional Lie algebras*, Geom. Dedicata **25** (1988), no. 1-3, 407–416, Geometries and groups (Noordwijkerhout, 1986).

[Ho] D.W. Hoffmann, *On the dimensions of anisotropic quadratic forms in I^4*, Invent. Math. **131** (1998), 185–198.

[HT] D.W. Hoffmann and J.-P. Tignol, *On 14-dimensional forms in I^3, 8-dimensional forms in I^2, and the common value property*, Doc. Math. **3** (1998), 189–214.

[Hu] J.E. Humphreys, *Introduction to Lie algebras and representation theory*, Graduate Texts in Mathematics, vol. 9, Springer-Verlag, 1980, Third printing, revised.

[Ig] J.-I. Igusa, *A classification of spinors up to dimension twelve*, Amer. J. Math. **92** (1970), 997–1028.

[IK] O.T. Izhboldin and N.A. Karpenko, *Some new examples in the theory of quadratic forms*, Math. Zeit. **234** (2000), 647–695.

[J 59] N. Jacobson, *Some groups of transformations defined by Jordan algebras. I*, J. Reine Angew. Math. **201** (1959), 178–195 (= Coll. Math. Papers 63).

[J 68] ———, *Structure and representations of Jordan algebras*, Coll. Pub., vol. 39, Amer. Math. Soc., Providence, RI, 1968.

[J 71] ———, *Exceptional Lie algebras*, Lecture Notes in Pure and Applied Mathematics, vol. 1, Marcel Dekker, New York, 1971.

[Ka] B. Kahn, *Comparison of some field invariants*, J. Algebra **232** (2000), 485–492.

[KMRT] M.-A. Knus, A.S. Merkurjev, M. Rost, and J.-P. Tignol, *The book of involutions*, Colloquium Publications, vol. 44, Amer. Math. Soc., 1998.

[Kn] M.-A. Knus, *Quadratic and hermitian forms over rings*, Grundlehren der mathematischen Wissenschaften, vol. 294, Springer, 1991.

[Kr] S. Krutelevich, *Jordan algebras, exceptional groups, and Bhargava composition*, J. Algebra **314** (2007), 924–977.

[Lam] T.Y. Lam, *Introduction to quadratic forms over fields*, Graduate Studies in Mathematics, vol. 67, Amer. Math. Soc., Providence, RI, 2005.

[Lang] S. Lang, *Topics in cohomology of groups*, Lecture Notes in Mathematics, vol. 1625, Springer, 1996.

[Li] M. Liebeck, *The affine permutation groups of rank three*, Proc. London Math. Soc. (3) **54** (1987), 477–516.

[Lu] J. Lurie, *On simply laced Lie algebras and their minuscule representations*, Comment. Math. Helv. **76** (2001), 515–575.

[MacD] M. MacDonald, PhD thesis in preparation, Cambridge University.

[McC] K. McCrimmon, *The Freudenthal-Springer-Tits constructions of exceptional Jordan algebras*, Trans. Amer. Math. Soc. **139** (1969), 495–510.

[McKP] W.G. McKay and J. Patera, *Tables of dimensions, indices, and branching rules for representations of simple Lie algebras*, Lecture Notes in Pure and Applied Mathematics, vol. 69, Marcel Dekker, New York, 1981.

[Mer] A.S. Merkurjev, *Rost invariants of simply connected algebraic groups*, with a section by S. Garibaldi, in [**GMS**].

[Mey] K. Meyberg, *Eine Theorie der Freudenthalschen Tripelsysteme. I, II*, Nederl. Akad. Wetensch. Proc. Ser. A 71 = Indag. Math. **30** (1968), 162–190.

[MPT] A.S. Merkurjev, R. Parimala, and J.-P. Tignol, *Invariants of quasi-trivial tori and the Rost invariant*, St. Petersburg Math. J. **14** (2003), 791–821.

[MPW] A.S. Merkurjev, I.A. Panin, A.R. Wadsworth, *Index reduction formulas for twisted flag varieties. II*, K-theory **14** (1998), 101–196.

[OVV] D. Orlov, A. Vishik, and V. Voevodsky, *An exact sequence for $K_*^M/2$ with applications to quadratic forms*, Ann. of Math. (2) **165** (2007), 1–13.

[Pf] A. Pfister, *Quadratische Formen in beliebigen Körpern*, Invent. Math. **1** (1966), 116–132.

[PeR 84] H.P. Petersson and M.L. Racine, *Springer forms and the first Tits construction of exceptional Jordan division algebras*, Manuscripta Math. **45** (1984), 249–272.

[PeR 94] ———, *Albert algebras*, Jordan algebras (Berlin) (W. Kaup, K. McCrimmon, and H.P. Petersson, eds.), de Gruyter, 1994, (Proceedings of a conference at Oberwolfach, 1992), pp. 197–207.

[PeR 96] ———, *An elementary approach to the Serre-Rost invariant of Albert algebras*, Indag. Math. (N.S.) **7** (1996), no. 3, 343–365.

[PlR] V.P. Platonov and A. Rapinchuk, *Algebraic groups and number theory*, Academic Press Inc., Boston, MA, 1994.

[Po] V.L. Popov, *Classification of spinors of dimension 14*, Trans. Moscow Math. Soc. (1980), 181–232.

[PoV] V.L. Popov and E.B. Vinberg, *Invariant theory*, Encyclopedia of Mathematical Sciences, vol. 55, pp. 123–284, Springer-Verlag, 1994.

[Pr] G. Prasad, *On the Kneser-Tits Problem for triality forms*, preprint, 2005.

[Re] I. Reiner, *Maximal orders*, LMS Monographs, vol. 5, Academic Press, 1975.

[Rö 93a] G. Röhrle, *On certain stabilizers in algebraic groups*, Comm. Algebra **21** (1993), no. 5, 1631–1644.

[Rö 93b] ———, *On extraspecial parabolic subgroups*, Contemp. Math., vol. 153, pp. 143–155, Amer. Math. Soc., Providence, RI, 1993.

[Rö 93c] ———, *On the structure of parabolic subgroups in algebraic groups*, J. Algebra **157** (1993), 80–115.

[Rost 91] M. Rost, *A (mod 3) invariant for exceptional Jordan algebras*, C. R. Acad. Sci. Paris Sér. I Math. **313** (1991), 823–827.

[Rost 99a] ———, *A descent property for Pfister forms*, J. Ramanujan Math. Soc. **14** (1999), no. 1, 55–63.

[Rost 99b] ———, *On 14-dimensional quadratic forms, their spinors, and the difference of two octonion algebras*, March 1999, unpublished manuscript available at `www.mathematik.uni-bielefeld.de/~rost/`

[Rost 99c] ———, *On the Galois cohomology of* Spin(14), original version March 1999, current version 3 June 2006, unpublished manuscript available at `www.mathematik.uni-bielefeld.de/~rost/`

[Rost 02] ———, *On the classification of Albert algebras*, September 2002, unpublished manuscript available at `www.mathematik.uni-bielefeld.de/~rost/`

[RST] M. Rost, J-P. Serre, and J.-P. Tignol, *La forme trace d'une algèbre simple centrale de degré 4*, C. R. Acad. Sci. Paris Sér. I Math. **342** (2006), 83–87.

[Ru] H. Rubenthaler, *Non-parabolic prehomogeneous vector spaces and exceptional Lie algebras*, J. Algebra **281** (2004), no. 1, 366–394.

[RY] Z. Reichstein and B. Youssin, *Essential dimensions of algebraic groups and a resolution theorem for G-varieties*, Canad. J. Math. **52** (2000), no. 5, 1018–1056, with an appendix by J. Kollár and E. Szabó.

[Sc] W. Scharlau, *Quadratic and hermitian forms*, Grundlehren der mathematischen Wissenschaften, vol. 270, Springer, 1985.

[Se 95] J-P. Serre, *Cohomologie galoisienne: progrès et problèmes*, Astérisque (1995), no. 227, 229–257, Séminaire Bourbaki, vol. 1993/94, Exp. 783, (= Oe. 166).

[Se 02] ———, *Galois cohomology*, Springer-Verlag, 2002, originally published as *Cohomologie galoisienne* (1965).

[SK] M. Sato and T. Kimura, *A classification of irreducible prehomogeneous vector spaces and their relative invariants*, Nagoya Math. J. **65** (1977), 1–155.

[Sl] N.J.A. Sloane, *The on-line encyclopedia of integer sequences*, published electronically at `www.research.att.com/~njas/sequences/`

[Sp 62] T.A. Springer, *Characterization of a class of cubic forms*, Nederl. Akad. Wetensch. **65** (1962), 259–265.

[Sp 98] ———, *Linear algebraic groups*, second ed., Birkhäuser, 1998.

[SS] T.A. Springer and R. Steinberg, *Conjugacy classes*, Seminar on Algebraic Groups and Related Finite Groups (The Institute for Advanced Study, Princeton, N.J., 1968/69), Springer, Berlin, 1970, pp. 167–266.

[St] R. Steinberg, *Lectures on Chevalley groups*, Yale University, New Haven, Conn., 1968.

[SV] T.A. Springer and F.D. Veldkamp, *Octonions, Jordan algebras and exceptional groups*, Springer-Verlag, Berlin, 2000.

[Th] M.L. Thakur, *Isotopy and invariants of Albert algebras*, Comm. Math. Helv. **74** (1999), 297–305.

[Tig] J.-P. Tignol, *La forme seconde trace d'une algèbre simple centrale de degré 4 de caractéristique 2*, C. R. Acad. Sci. Paris Sér. I Math. **342** (2006), 89–92.

[Tits 66] J. Tits, *Classification of algebraic semisimple groups*, Algebraic Groups and Discontinuous Subgroups, Proc. Symp. Pure Math., vol. IX, AMS, 1966, pp. 32–62.

[Tits 71] ———, *Représentations linéaires irréductibles d'un groupe réductif sur un corps quelconque*, J. Reine Angew. Math. **247** (1971), 196–220.

[Tits 90] ———, *Strongly inner anisotropic forms of simple algebraic groups*, J. Algebra **131** (1990), 648–677.

[Tits 92] ———, *Sur les degrés des extensions de corps déployant les groupes algébriques simples*, C. R. Acad. Sci. Paris Sér. I Math. **315** (1992), no. 11, 1131–1138.

[To 04] B. Totaro, *Splitting fields for E_8-torsors*, Duke Math. J. **121** (2004), 425–455.

[To 05] ———, *The torsion index of E_8 and other groups*, Duke Math. J. **129** (2005), 219–248.

[Va] V.S. Varadarajan, Spin(7)*-subgroups of* SO(8) *and* Spin(8), Expo. Math. **19** (2001), 163–177.

[Vi] E.B. Vinberg, *The Weyl group of a graded Lie algebra*, Math. USSR Izv. **10** (1976), 463–495.

[Vo] V. Voevodsky, *Motivic cohomology with $\mathbb{Z}/2$-coefficients*, Inst. Hautes Études Sci. Publ. Math. **98** (2003), no. 1, 59–104.

[Wa] W.C. Waterhouse, *Discriminants of étale algebras and related structures*, J. Reine Angew. Math. **379** (1987), 209–220.

[We] E. Weiss, *Cohomology of groups*, Academic Press, 1969.

[WS] A.R. Wadsworth and D.B. Shapiro, *On multiples of round and Pfister forms*, Math. Zeit. **157** (1977), 53–62.

Index

Editorial Information

To be published in the *Memoirs*, a paper must be correct, new, nontrivial, and significant. Further, it must be well written and of interest to a substantial number of mathematicians. Piecemeal results, such as an inconclusive step toward an unproved major theorem or a minor variation on a known result, are in general not acceptable for publication.

Papers appearing in *Memoirs* are generally at least 80 and not more than 200 published pages in length. Papers less than 80 or more than 200 published pages require the approval of the Managing Editor of the Transactions/Memoirs Editorial Board.

As of March 31, 2009, the backlog for this journal was approximately 12 volumes. This estimate is the result of dividing the number of manuscripts for this journal in the Providence office that have not yet gone to the printer on the above date by the average number of monographs per volume over the previous twelve months, reduced by the number of volumes published in four months (the time necessary for preparing a volume for the printer). (There are 6 volumes per year, each usually containing at least 4 numbers.)

A Consent to Publish and Copyright Agreement is required before a paper will be published in the *Memoirs*. After a paper is accepted for publication, the Providence office will send a Consent to Publish and Copyright Agreement to all authors of the paper. By submitting a paper to the *Memoirs*, authors certify that the results have not been submitted to nor are they under consideration for publication by another journal, conference proceedings, or similar publication.

Information for Authors

Memoirs are printed from camera copy fully prepared by the author. This means that the finished book will look exactly like the copy submitted.

Initial submission. The AMS uses Centralized Manuscript Processing for initial submissions. Authors should submit a PDF file using the Initial Manuscript Submission form found at `www.ams.org/peer-review-submission`, or send one copy of the manuscript to the following address: Centralized Manuscript Processing, MEMOIRS OF THE AMS, 201 Charles Street, Providence, RI 02904-2294 USA. If a paper copy is being forwarded to the AMS, indicate that it is for it Memoirs and include the name of the corresponding author, contact information such as email address or mailing address, and the name of an appropriate Editor to review the paper (see the list of Editors below).

The paper must contain a *descriptive title* and an *abstract* that summarizes the article in language suitable for workers in the general field (algebra, analysis, etc.). The *descriptive title* should be short, but informative; useless or vague phrases such as "some remarks about" or "concerning" should be avoided. The *abstract* should be at least one complete sentence, and at most 300 words. Included with the footnotes to the paper should be the 2000 *Mathematics Subject Classification* representing the primary and secondary subjects of the article. The classifications are accessible from `www.ams.org/msc/`. The list of classifications is also available in print starting with the 1999 annual index of *Mathematical Reviews*. The Mathematics Subject Classification footnote may be followed by a list of *key words and phrases* describing the subject matter of the article and taken from it. Journal abbreviations used in bibliographies are listed in the latest *Mathematical Reviews* annual index. The series abbreviations are also accessible from `www.ams.org/msnhtml/serials.pdf`. To help in preparing and verifying references, the AMS offers MR Lookup, a Reference Tool for Linking, at `www.ams.org/mrlookup/`.

Electronically prepared manuscripts. The AMS encourages electronically prepared manuscripts, with a strong preference for $\mathcal{A}_{\mathcal{M}}\mathcal{S}$-LaTeX. To this end, the Society has prepared $\mathcal{A}_{\mathcal{M}}\mathcal{S}$-LaTeX author packages for each AMS publication. Author packages include instructions for preparing electronic manuscripts, samples, and a style file that generates the particular design specifications of that publication series. Though $\mathcal{A}\mathcal{M}\mathcal{S}$-LaTeX is the highly preferred format of TeX, author packages are also available in $\mathcal{A}\mathcal{M}\mathcal{S}$-TeX.

Authors may retrieve an author package for *Memoirs of the AMS* from `www.ams.org/journals/memo/memoauthorpac.html` or via FTP to `ftp.ams.org` (login as `anonymous`, enter username as password, and type `cd pub/author-info`). The *AMS Author Handbook* and the *Instruction Manual* are available in PDF format from the author package link. The author package can also be obtained free of charge by sending email to `tech-support@ams.org` (Internet) or from the Publication Division, American Mathematical Society, 201 Charles St., Providence, RI 02904-2294, USA. When requesting an author package, please specify $\mathcal{A}\mathcal{M}\mathcal{S}$-LaTeX or $\mathcal{A}\mathcal{M}\mathcal{S}$-TeX and the publication in which your paper will appear. Please be sure to include your complete mailing address.

After acceptance. The final version of the electronic file should be sent to the Providence office (this includes any TeX source file, any graphics files, and the DVI or PostScript file) immediately after the paper has been accepted for publication.

Before sending the source file, be sure you have proofread your paper carefully. The files you send must be the EXACT files used to generate the proof copy that was accepted for publication. For all publications, authors are required to send a printed copy of their paper, which exactly matches the copy approved for publication, along with any graphics that will appear in the paper.

Accepted electronically prepared files can be submitted via the web at `www.ams.org/submit-book-journal/`, sent via FTP, or sent on CD-Rom or diskette to the Electronic Prepress Department, American Mathematical Society, 201 Charles Street, Providence, RI 02904-2294 USA. TeX source files, DVI files, and PostScript files can be transferred over the Internet by FTP to the Internet node `ftp.ams.org` (130.44.1.100). When sending a manuscript electronically via CD-Rom or diskette, please be sure to include a message identifying the paper as a Memoir.

Electronically prepared manuscripts can also be sent via email to `pub-submit@ams.org` (Internet). In order to send files via email, they must be encoded properly. (DVI files are binary and PostScript files tend to be very large.)

Electronic graphics. Comprehensive instructions on preparing graphics are available at `www.ams.org/authors/journals.html`. A few of the major requirements are given here.

Submit files for graphics as EPS (Encapsulated PostScript) files. This includes graphics originated via a graphics application as well as scanned photographs or other computer-generated images. If this is not possible, TIFF files are acceptable as long as they can be opened in Adobe Photoshop or Illustrator. No matter what method was used to produce the graphic, it is necessary to provide a paper copy to the AMS.

Authors using graphics packages for the creation of electronic art should also avoid the use of any lines thinner than 0.5 points in width. Many graphics packages allow the user to specify a "hairline" for a very thin line. Hairlines often look acceptable when proofed on a typical laser printer. However, when produced on a high-resolution laser imagesetter, hairlines become nearly invisible and will be lost entirely in the final printing process.

Screens should be set to values between 15% and 85%. Screens which fall outside of this range are too light or too dark to print correctly. Variations of screens within a graphic should be no less than 10%.

Inquiries. Any inquiries concerning a paper that has been accepted for publication should be sent to `memo-query@ams.org` or directly to the Electronic Prepress Department, American Mathematical Society, 201 Charles St., Providence, RI 02904-2294 USA.

Editors

This journal is designed particularly for long research papers, normally at least 80 pages in length, and groups of cognate papers in pure and applied mathematics. Papers intended for publication in the *Memoirs* should be addressed to one of the following editors. The AMS uses Centralized Manuscript Processing for initial submissions to AMS journals. Authors should follow instructions listed on the Initial Submission page found at www.ams.org/memo/memosubmit.html.

Algebra to ALEXANDER KLESHCHEV, Department of Mathematics, University of Oregon, Eugene, OR 97403-1222; email: ams@noether.uoregon.edu

Algebraic geometry to DAN ABRAMOVICH, Department of Mathematics, Brown University, Box 1917, Providence, RI 02912; email: amsedit@math.brown.edu

Algebraic geometry and its applications to MINA TEICHER, Emmy Noether Research Institute for Mathematics, Bar-Ilan University, Ramat-Gan 52900, Israel; email: teicher@macs.biu.ac.il

Algebraic topology to ALEJANDRO ADEM, Department of Mathematics, University of British Columbia, Room 121, 1984 Mathematics Road, Vancouver, British Columbia, Canada V6T 1Z2; email: adem@math.ubc.ca

Combinatorics to JOHN R. STEMBRIDGE, Department of Mathematics, University of Michigan, Ann Arbor, Michigan 48109-1109; email: JRS@umich.edu

Commutative and homological algebra to LUCHEZAR L. AVRAMOV, Department of Mathematics, University of Nebraska, Lincoln, NE 68588-0130; email: avramov@math.unl.edu

Complex analysis and harmonic analysis to ALEXANDER NAGEL, Department of Mathematics, University of Wisconsin, 480 Lincoln Drive, Madison, WI 53706-1313; email: nagel@math.wisc.edu

Differential geometry and global analysis to CHRIS WOODWARD, Department of Mathematics, Rutgers University, 110 Frelinghuysen Road, Piscataway, NJ 08854; email: ctw@math.rutgers.edu

Dynamical systems and ergodic theory and complex analysis to YUNPING JIANG, Department of Mathematics, CUNY Queens College and Graduate Center, 65-30 Kissena Blvd., Flushing, NY 11367; email: Yunping.Jiang@qc.cuny.edu

Functional analysis and operator algebras to DIMITRI SHLYAKHTENKO, Department of Mathematics, University of California, Los Angeles, CA 90095; email: shlyakht@math.ucla.edu

Geometric analysis to WILLIAM P. MINICOZZI II, Department of Mathematics, Johns Hopkins University, 3400 N. Charles St., Baltimore, MD 21218; email: trans@math.jhu.edu

Geometric topology to MARK FEIGHN, Math Department, Rutgers University, Newark, NJ 07102; email: feighn@andromeda.rutgers.edu

Harmonic analysis, representation theory, and Lie theory to ROBERT J. STANTON, Department of Mathematics, The Ohio State University, 231 West 18th Avenue, Columbus, OH 43210-1174; email: stanton@math.ohio-state.edu

Logic to STEFFEN LEMPP, Department of Mathematics, University of Wisconsin, 480 Lincoln Drive, Madison, Wisconsin 53706-1388; email: lempp@math.wisc.edu

Number theory to JONATHAN ROGAWSKI, Department of Mathematics, University of California, Los Angeles, CA 90095; email: jonr@math.ucla.edu

Number theory to SHANKAR SEN, Department of Mathematics, 505 Malott Hall, Cornell University, Ithaca, NY 14853; email: ss70@cornell.edu

Partial differential equations to GUSTAVO PONCE, Department of Mathematics, South Hall, Room 6607, University of California, Santa Barbara, CA 93106; email: ponce@math.ucsb.edu

Partial differential equations and dynamical systems to PETER POLACIK, School of Mathematics, University of Minnesota, Minneapolis, MN 55455; email: polacik@math.umn.edu

Probability and statistics to RICHARD BASS, Department of Mathematics, University of Connecticut, Storrs, CT 06269-3009; email: bass@math.uconn.edu

Real analysis and partial differential equations to DANIEL TATARU, Department of Mathematics, University of California, Berkeley, Berkeley, CA 94720; email: tataru@math.berkeley.edu

All other communications to the editors should be addressed to the Managing Editor, ROBERT GURALNICK, Department of Mathematics, University of Southern California, Los Angeles, CA 90089-1113; email: guralnic@math.usc.edu.

Titles in This Series

941 **Gelu Popescu,** Unitary invariants in multivariable operator theory, 2009

940 **Gérard Iooss and Pavel I. Plotnikov,** Small divisor problem in the theory of three-dimensional water gravity waves, 2009

939 **I. D. Suprunenko,** The minimal polynomials of unipotent elements in irreducible representations of the classical groups in odd characteristic, 2009

938 **Antonino Morassi and Edi Rosset,** Uniqueness and stability in determining a rigid inclusion in an elastic body, 2009

937 **Skip Garibaldi,** Cohomological invariants: Exceptional groups and spin groups, 2009

936 **André Martinez and Vania Sordoni,** Twisted pseudodifferential calculus and application to the quantum evolution of molecules, 2009

935 **Mihai Ciucu,** The scaling limit of the correlation of holes on the triangular lattice with periodic boundary conditions, 2009

934 **Arjen Doelman, Björn Sandstede, Arnd Scheel, and Guido Schneider,** The dynamics of modulated wave trains, 2009

933 **Luchezar Stoyanov,** Scattering resonances for several small convex bodies and the Lax-Phillips conjuecture, 2009

932 **Jun Kigami,** Volume doubling measures and heat kernel estimates of self-similar sets, 2009

931 **Robert C. Dalang and Marta Sanz-Solé,** Hölder-Sobolv regularity of the solution to the stochastic wave equation in dimension three, 2009

930 **Volkmar Liebscher,** Random sets and invariants for (type II) continuous tensor product systems of Hilbert spaces, 2009

929 **Richard F. Bass, Xia Chen, and Jay Rosen,** Moderate deviations for the range of planar random walks, 2009

928 **Ulrich Bunke,** Index theory, eta forms, and Deligne cohomology, 2009

927 **N. Chernov and D. Dolgopyat,** Brownian Brownian motion-I, 2009

926 **Riccardo Benedetti and Francesco Bonsante,** Canonical wick rotations in 3-dimensional gravity, 2009

925 **Sergey Zelik and Alexander Mielke,** Multi-pulse evolution and space-time chaos in dissipative systems, 2009

924 **Pierre-Emmanuel Caprace,** "Abstract" homomorphisms of split Kac-Moody groups, 2009

923 **Michael Jöllenbeck and Volkmar Welker,** Minimal resolutions via algebraic discrete Morse theory, 2009

922 **Ph. Barbe and W. P. McCormick,** Asymptotic expansions for infinite weighted convolutions of heavy tail distributions and applications, 2009

921 **Thomas Lehmkuhl,** Compactification of the Drinfeld modular surfaces, 2009

920 **Georgia Benkart, Thomas Gregory, and Alexander Premet,** The recognition theorem for graded Lie algebras in prime characteristic, 2009

919 **Roelof W. Bruggeman and Roberto J. Miatello,** Sum formula for SL_2 over a totally real number field, 2009

918 **Jonathan Brundan and Alexander Kleshchev,** Representations of shifted Yangians and finite W-algebras, 2008

917 **Salah-Eldin A. Mohammed, Tusheng Zhang, and Huaizhong Zhao,** The stable manifold theorem for semilinear stochastic evolution equations and stochastic partial differential equations, 2008

916 **Yoshikata Kida,** The mapping class group from the viewpoint of measure equivalence theory, 2008

TITLES IN THIS SERIES

915 **Sergiu Aizicovici, Nikolaos S. Papageorgiou, and Vasile Staicu,** Degree theory for operators of monotone type and nonlinear elliptic equations with inequality constraints, 2008

914 **E. Shargorodsky and J. F. Toland,** Bernoulli free-boundary problems, 2008

913 **Ethan Akin, Joseph Auslander, and Eli Glasner,** The topological dynamics of Ellis actions, 2008

912 **Igor Chueshov and Irena Lasiecka,** Long-time behavior of second order evolution equations with nonlinear damping, 2008

911 **John Locker,** Eigenvalues and completeness for regular and simply irregular two-point differential operators, 2008

910 **Joel Friedman,** A proof of Alon's second eigenvalue conjecture and related problems, 2008

909 **Cameron McA. Gordon and Ying-Qing Wu,** Toroidal Dehn fillings on hyperbolic 3-manifolds, 2008

908 **J.-L. Waldspurger,** L'endoscopie tordue n'est pas si tordue, 2008

907 **Yuanhua Wang and Fei Xu,** Spinor genera in characteristic 2, 2008

906 **Raphaël S. Ponge,** Heisenberg calculus and spectral theory of hypoelliptic operators on Heisenberg manifolds, 2008

905 **Dominic Verity,** Complicial sets characterising the simplicial nerves of strict ω-categories, 2008

904 **William M. Goldman and Eugene Z. Xia,** Rank one Higgs bundles and representations of fundamental groups of Riemann surfaces, 2008

903 **Gail Letzter,** Invariant differential operators for quantum symmetric spaces, 2008

902 **Bertrand Toën and Gabriele Vezzosi,** Homotopical algebraic geometry II: Geometric stacks and applications, 2008

901 **Ron Donagi and Tony Pantev (with an appendix by Dmitry Arinkin),** Torus fibrations, gerbes, and duality, 2008

900 **Wolfgang Bertram,** Differential geometry, Lie groups and symmetric spaces over general base fields and rings, 2008

899 **Piotr Hajłasz, Tadeusz Iwaniec, Jan Malý, and Jani Onninen,** Weakly differentiable mappings between manifolds, 2008

898 **John Rognes,** Galois extensions of structured ring spectra/Stably dualizable groups, 2008

897 **Michael I. Ganzburg,** Limit theorems of polynomial approximation with exponential weights, 2008

896 **Michael Kapovich, Bernhard Leeb, and John J. Millson,** The generalized triangle inequalities in symmetric spaces and buildings with applications to algebra, 2008

895 **Steffen Roch,** Finite sections of band-dominated operators, 2008

894 **Martin Dindoš,** Hardy spaces and potential theory on C^1 domains in Riemannian manifolds, 2008

893 **Tadeusz Iwaniec and Gaven Martin,** The Beltrami Equation, 2008

892 **Jim Agler, John Harland, and Benjamin J. Raphael,** Classical function theory, operator dilation theory, and machine computation on multiply-connected domains, 2008

891 **John H. Hubbard and Peter Papadopol,** Newton's method applied to two quadratic equations in $\mathbb{C}^2$ viewed as a global dynamical system, 2008

890 **Steven Dale Cutkosky,** Toroidalization of dominant morphisms of 3-folds, 2007

889 **Michael Sever,** Distribution solutions of nonlinear systems of conservation laws, 2007

For a complete list of titles in this series, visit the AMS Bookstore at **www.ams.org/bookstore/**.